Small-scale Cogeneration Handbook

Second Edition

SMALL-SCALE COGENERATION HANDBOOK
SECOND EDITION

Bernard F. Kolanowski, BSME

THE FAIRMONT PRESS, INC.
Lilburn, Georgia

MARCEL DEKKER, INC.
New York and Basel

Library of Congress Cataloging-in-Publication Data

Kolanowski, Bernard F.
    Small-scale cogeneration handbook/Bernard F. Kolanowski.—2nd ed.
        p. cm.
    ISBN 0-88173-419-5 (print)
    1. Cogeneration of electric power and heat. I. Title.

TK1041.K58 2003
621.1'99--dc21   2002192853

Small-scale cogeneration handbook by Bernard F. Kolanowski--Second Edition
©2003 by The Fairmont Press. All rights reserved. No part of this publication may be reproduced or transmitted in any form or by any means, electronic or mechanical, including photocopy, recording, or any information storage and retrieval system, without permission in writing from the publisher.

Published by The Fairmont Press, Inc.
700 Indian Trail, Lilburn, GA 30047
tel: 770-925-9388; fax: 770-381-9865
http://fairmontpress.com

Distributed by Marcel Dekker, Inc.
270 Madison Avenue, New York, NY 10016
tel: 212-696-9000; fax: 212-685-4540
http://www.dekker.com

Printed in the United States of America

10 9 8 7 6 5 4 3 2 1

0-88173-419-5 (The Fairmont Press, Inc.)
0-8247-4122-6 (Marcel Dekker, Inc.)

While every effort is made to provide dependable information, the publisher, authors, and editors cannot be held responsible for any errors or omissions.

Dedication

This book is dedicated to the people responsible for giving me the opportunity to get intimately involved in the field of small-scale cogeneration: Mr. Herbert Ratch (deceased), Mr. David Lumbert, and Ms. Margo (Ratch) Bennett.

and

To my loving wife, Mary Beth

Table of Contents

Chapter		Page
1	Introduction	1
2	History of Cogeneration	5
3	Regulatory Actions	11
4	Uses of Cogeneration	21
5	Applying Cogeneration	25
6	Sizing the Cogenerator	29
7	Logistics of Installation	45
8	Permitting Requirements	51
9	Operation & Maintenance	59
10	Pitfalls of Cogeneration	65
11	Financing Cogeneration Projects	73
12	Case Histories	79
13	Small-scale Cogeneration Manufacturers	87
14	Do-it-yourself Cogeneration	97
15	Green Energy vs. Cogeneration	107
16	Micro-turbines and Cogenration	111
17	Absorber Chillers in Cogeneration	123
18	Distributed Generation	129
19	National Combined Heat & Power Association	137
20	Cogeneration in Europe	149

Appendices

I	Typical Shared Savings Agreement	157
II	Energy Conversion Tables	169
III	Heat Loss in Swimming Pools—Graphs	170
IV	Typical Small-scale Cogeneration Energy Production	172
V	Typical Utility Rate Schedules—Gas and Electricity	173
VI	Working a Cogeneration Project—A Model	197

Index 203

Preface

Whenever you turn on the heater in your car, you are cogenerating. The heat your engine would normally dissipate through your radiator is passed through your car's heater warming the inside of the passenger compartment as you drive along. The single fuel you are using is the gasoline (or natural gas) in your tank, but it is providing both the power to drive your car and the heat to keep you warm.

And that's what cogeneration is. Using one fuel to produce two usable energy sources. In this discussion the fuel will be natural gas (although propane and diesel oil may also be used) and the usable energy will be electricity and hot water.

WHY COGENERATE?

You will use cogeneration to save money, but there are other benefits of cogeneration if you are concerned with the air you breathe and the energy used from mother earth.

Money; because cogeneration produces two usable energy sources from a single fuel, it operates more efficiently than your present sources of energy. In fact, cogeneration turns up to 90% of the fuel burned into usable energy. That compares with just 52% of the fuel burned in the local power plant and in your existing hot water heater.

That difference in efficiency saves money. Returns on investment of capital range from 25 to 50% when cogeneration is properly applied to a facility, be it a home, commercial operation or an industry. Cogeneration will throw off a positive cash flow after paying for the residual energy you buy from the utility, the debt service for the investment and the maintenance and operating costs.

Environmental, because cogeneration uses less fuel overall, less pollutants will be emitted to the atmosphere. Even in strict Air Quality Districts catalytic converters, similar to those in your automobile, will protect the atmosphere from excess pollution.

Conservational, because cogeneration will burn less fuel, the energy resources of this planet will be conserved.

WHO IS COGENERATING?

You'd be surprised at the variety of commercial and industrial businesses that are using cogeneration to cut costs and conserve energy and the environment:

- Did you visit your local fitness center this week? They are a natural for cogeneration because of the hot water they use in spas, swimming pools and showers.

- Wastewater treatment plants use cogeneration by burning a combination of biogas and natural gas to generate electricity and hot water.

- Municipalities use cogeneration to heat swimming pools and air condition associated meeting and fitness centers.

- Food processors use cogeneration to cook and pasteurize their products while generating electricity for internal consumption.

- Casinos are using cogeneration in conjunction with absorber-chillers to cool and electrify their facilities.

- Hotels are using cogeneration for guest room hot water, laundry hot water, kitchen hot water as well as heating their pools and spas while generating a percentage of their electrical needs.

- Hospitals are cogenerating for heating therapy pools to over 90 degrees, sterilizing operating instruments, and general hot water needs while making most of their electrical needs.

WHERE DO YOU COGENERATE?

On your site. The typical cogeneration system is no bigger than an executive size desk, and just about as quiet, too. Placed on site, it ties in with the existing hot water heating system and electrical distribution system to provide the facility with the first line in heating water and electrifying the facility. When more hot water or electricity is needed than the cogenerator can supply, the existing systems kick in and furnish that excess without missing a beat. When the cogenerator needs maintenance, those same existing systems that used to supply all of your utility needs are there to insure your facility keeps operating. Cogenerators can even be engineered to continue to operate during central utility power failures if that is important to you.

CAN YOU FINANCE THE COGENERATION SYSTEM?

Yes! Conventional financing through your bank or lending institution is one way. State assisted financing for energy conservation projects is available in many states. Grant money from both federal and state governments is another way. Leasing is a very popular way to do off-balance-sheet financing. Another way to attain the benefits of cogeneration and have no capital outlay is to utilize third party financing via Shared Savings Agreements.

SHOULD YOU COGENERATE?

Only if you want to save money and be a good neighbor in conserving energy and combating pollution.

This book presents the state of the art and science of the technology of cogeneration while demonstrating the practical side of implementing this art and overcoming the pitfalls while staying within the changing regulatory boundaries required to bring home a successful cogeneration project.

Acknowledgments

The following companies and/or people have contributed to this book:

Intelligent Solutions, Inc.
Capstone Turbine Corporation
Yazaki Corporation
San Diego Gas & Electric Co.
Hannes Hunschofsky
Trigen-Ewing Corporation
EPRI
Broad, USA
Ingersoll-Rand Corp.
Tecogen
Turbec Americas
Bowman Power Systems

Chapter 1

Introduction

Some years ago as I was moving into a new neighborhood I met some of the people I would be living around. One neighbor asked what I do for a living and I replied, "I'm in cogeneration."

He said, "What is that?" After explaining he then said, "Next time someone asks you what you do, tell them you're in real estate. Everyone knows what real estate is!"

And that's the way it's been ever since. About one in ten people will have any idea of what cogeneration is, and even that one will have a somewhat glazed look to his eyes. So, to explain to both the readers who are learning about cogeneration as well as those that have experience in this 'exotic' field I will elaborate. Cogeneration is the simultaneous production of two or more beneficial work outputs from a singular source of fuel input. In small-scale cogeneration the two work outputs are almost exclusively electricity and hot water and the single fuel input is natural gas.

An example to which most people can relate is the automobile. When automobiles were first introduced few, if any, had a heater built into the car. To enjoy both the thrill of powering yourself down the road and interior comfort during cold weather, one might have a kerosene heater inside the car to provide heating comfort. Therefore, when you stopped at the gas station you would buy gasoline for the car's engine and kerosene for the internal heater. You were getting two work outputs—motive power and interior comfort heating—but you were using two different fuels. No cogeneration here.

Then, when automobile manufacturers decided to place heaters inside the car rather than using a heater with a separate fuel source, they recognized the fact that the engine was throwing off a vast quantity of heat through its radiator. They devised a method whereby some of this heat could be channeled into the car's interior for the comfort of its driver and passengers, This was truly cogeneration: two work outputs from a single fuel input.

So in the case of small-scale cogeneration, which is the primary topic of this book, it will be shown how electricity and hot water can be created from a single fuel source and the savings that accrue from this rather simple technology. When you get a fuel to do two work efforts the savings will be in fuel costs. Electricity is generated from a prime mover, most often a reciprocating engine, driving an electric generator. The waste heat from the engine is channeled through heat exchangers to heat water that would normally have been heated by a separate hot water heater in the facility. The engine heat is captured primarily in the engine's coolant, an ethylene glycol antifreeze solution, and that coolant is pumped to a heat exchanger to transfer its heat to the water needing to be heated. The ethylene glycol loses some of its heat to the water and is circulated back to the engine block to be reheated by the engine and pumped back to the hot water heat exchanger. Since there is also heat in the engine's exhaust that heat is captured in an exhaust gas heat exchanger which transfers this waste heat to the ethylene glycol. A typical system is shown in Figure 1.

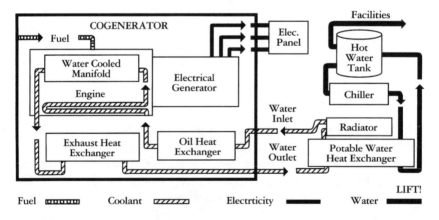

Figure 1-1

The efficiency of this system in converting the energy in the fuel to useful work is quite high. Using data from one of the manufacturers of small-scale cogenerators, it can be shown that a system designed to produce 120 kilowatts (kW) of electricity and 5.62 therms of thermal energy (hot water) has a fuel-usage efficiency of more than 90%. The fuel input is 10.7 therms of natural gas, or 1,070,000 Btu's of energy input. One hundred twenty kW of electricity @ 3415 Btu's/kW = 409,800 Btu's of energy output, plus the 562,000 Btu's of energy output in the form of hot water. Therefore, a total of 409,800 + 562,000 = 971,800 Btu's of energy output vs. 1,070,000 Btu's of energy input. The thermal efficiency is 971,800/1,070,000 = 90.82%.

The electricity produced at the central station has an overall efficiency of about 36% delivered to the customer's facility. That takes into account not only the fuel used to create the electricity, but also the transmission losses incurred in getting that electricity to the facility. A typical on-site hot-water heater has an efficiency of 75 to 80%. So, to deliver 120 kW of electricity, the central station utility must burn 120 kW × 3415 Btu/kW/0.36 = 1,138,333 Btu's of fuel. The hot water heater must burn 562,000 Btu's/0.75 = 749,333 Btu's. That's a total of 1,887,666 Btu's of fuel burned to create 971,800 Btu's of useful work. That's an efficiency of 971,800/1,887,666 = 51.48%.

A customer buying electricity from a central-station utility and heating water in his on-site water heaters will purchase 817,666 Btu's more fuel to gain the same useful energy than if he were cogenerating on site. And that is for EVERY HOUR HE NEEDS THAT ENERGY! A facility open seven days a week for 16 hours a day will buy 47,751 more therms of energy per year than the same facility who uses on-site cogeneration. At an average street cost of $0.75 per therm that's $35,813 more dollars spent just in fuel costs alone. However, as we shall see in later chapters, that's not the only costs involved since the utility, which buys fuel at a considerably lower cost than street costs, must add much more to the cost of electricity than simply the fuel costs.

The actual costs to the user in this example to purchase 120 kW of electricity and 562,000 Btu's of hot water for 7 days a week, 16 hours a day is over $80,000 per year.

Chapter 4 explores fully the economics of using cogeneration on site, but suffice it to say that the amortization of on-site cogeneration, i.e. the time to recoup the capital costs of the system, is an average of three years or less, even after accounting for operating and maintenance costs of the system.

Not only are economics involved in using cogeneration, but there are environmental and conservational issues that also benefit from getting more work with less fuel. Fifty-six percent less fuel burned will create 56% less pollutants in the atmosphere. Fifty-six percent less fuel burned will be 56% more fuel available for future generations of energy users.

Chapter 2

History of Cogeneration

The practical use of cogeneration is as old as the generation of electricity itself. When electrification of broad areas was devised to replace gas and kerosene lighting in residences and commercial facilities the concept of central station power generation plants was born. District heating systems were popular during the late 1800's and why not. District heating dates back to Roman times when warm water was circulated in open trenches to heat buildings and communal baths. District electrification dates back to Thomas Edison's plants in New York, and it didn't take long to combine the two. The prime movers that drove electric generators throw off waste heat that is normally blown to atmosphere. By capturing that heat and making low-pressure steam, that steam could be piped throughout the district for heating homes and businesses. Thus, cogeneration on a fairly large scale was born.

As electrification marched across the country, most of the generated electricity was on site in large industrial plants. With that generation, there is no doubt that much waste heat was captured and utilized in industrial processes as a natural offshoot. Probably the word cogeneration was not even used in conjunction with those efforts, but cogeneration it was. As large, central generating stations were built, it became cheaper for those industries that had been self-generating electricity to now buy from the central utility. With that change came the end to "cogeneration" in those industrial plants. Central station utility plants were now located off the beaten path, so even district heating suffered as the lines to connect to districts became too long and costly. Cheap oil and natural gas were the cause of our return to wastefulness, and little thought was given to energy efficiency when oil was selling for under a dollar a barrel.

But nothing is steady. Change is everything. With the first OPEC energy crisis in 1973 came a realization that America was no longer self-sufficient in supplying its total energy needs and that foreign countries now controlled what the price of energy would be. The oil produced in

America still only cost $4.00 per barrel, but if OPEC was going to sell its oil at $20.00 per barrel and we had to import over half our needs, then all oil was going to sell at the going rate. With expensive energy came plans to conserve energy and to seek energy supplies that were heretofore costly to get at and to seek alternative sources of energy. The famous tar sands of the Athabasca region in west central Canada were exploited when the cost of oil was predicted to go to $40 a barrel. Drilling rigs were punching holes all over the traditional oil-bearing areas of the United States opening small "stripper" wells and re-opening wells that had been abandoned due to the higher costs of production. America was in an oil boom only to see it burst when OPEC, knowing they controlled these matters, let the cost of oil slide to $10 a barrel, and we saw those efforts at exploiting domestic sources go wanting.

Conservation was now a household word. With the cost of electricity tied to the price of oil, consumers felt the pinch of rising electricity prices. An enterprising group of neighbors in the Bronx section of New York decided to put up a windmill to generate enough electricity to help cut their costs from Consolidated Edison, the major supplier of electricity in New York. They would still be tied into Con Ed's system, but when the wind blew, they could count on their costs being lowered by their wind-powered production of electricity. The system was so successful that at its peak it generated slightly more electricity than was needed at any given time, so they decided to sell this excess back to Con Ed, who, of course, was getting it free whenever an excess was generated. Con Ed balked at having to buy power from this upstart neighborhood and abjectly refused. The neighbors sued and won. From this meager beginning came the Public Utility Regulated Policy Act that we fondly call PURPA today.

The PURPA law paved the way for larger-scale cogeneration and independent power generation.

Very few businesses could afford to generate their own electricity exclusively. Variations in their power needs on an hour-by-hour basis; reliability and maintenance of the on-site generators; additions to their operations; all required the back up of the central station utility to make these independent power generators and cogenerators feasible. In effect, PURPA said that a central-station utility must allow interconnection of these facilities with their grid to act as standby and makeup power sources. It further said that the cost of the fuel to power these cogenerators would be similar to that which the central station utilities paid for their fuel.

Furthermore, it reinforced the law requiring the central station to purchase any excess power generated by these independent facilities at the "avoided cost" of the utility. The term "avoided cost" led to very creative accounting by the utilities to determine exactly what their "avoided cost" was. Too low, and their guaranteed return on investment would be jeopardized; too high and their payment for purchased kilowatts would be too expensive. It is doubtful that any two utilities in the country had identical policies when it came to determining their "avoided cost." The utilities that were selling kilowatts at 5 cents each were virtually immune to independent power producers and cogenerators, while those whose prices were 16 cents a kilowatt were now inundated with alternative sources of electricity. Abuses were rampant on both sides. Facilities were built in these high-priced regions, presumably as cogeneration facilities, but they were mostly power generators that wanted to capitalize on the cheap fuel costs for cogeneration systems and the other PURPA law advantages. When the utility was forced to pay 8 and 10 cents a kilowatt for "excess" power it became advantageous to produce "excess" power.

With these abuses came regulations. The Federal Energy Regulatory Commission, FERC, was set up by the government to put some ethics into the business of generating, buying, and selling independent power. An efficiency standard was set up requiring a cogeneration system to meet a minimum standard of thermal-energy utilization in order to derive the full benefits of PURPA.

The formula used is: (All inputs are in Btu's)

$$\text{Efficiency} = \frac{\text{Thermal Energy Produced}/2 + \text{Electrical Energy Produced}}{\text{Fuel Input}}$$

The minimum efficiency required to meet this formula is 42.5%.

Squabbles between utility and cogenerators still ensued, however, as the responsibility to prove FERC efficiency was argued. The cogenerators said that if the utilities wanted to know FERC efficiency, they could instrument and monitor the units. The utilities said it was up to the cogenerators to prove they were meeting the minimum efficiency levels. Cogenerators still wanted to generate as much electricity as possible as this was the motive force that paid for their investment. If the heat could be used, fine. But if it could not be used, it was then discarded via blow-off radiators while the electrical generation went on. These blow-off radiators, also called dump radiators, were integral to the system and were actuated thermally when the cogeneration heat transfer fluid, usually the engine coolant, reached a maximum temperature. To continue to operate the cogenerator with reduced heat transfer caused excess coolant temperatures, and the engine would shut down on high temperature.

Obviously, this would interrupt the on-site generation of electricity, and the engine could not be restarted until the temperature of the coolant came down. Not being practical, nor economical, the dump radiator was incorporated into the system to take care of these periods of little useful heat transfer. The utilities could not prove that the FERC efficiency was being violated; so if the cogenerators did not supply the information, the utility enforced the PURPA rule, and all gas that was purchased at the cogeneration rate would now be priced at the commercial rate and was retroactive to the previous three years. This was a hefty penalty. Even a small cogeneration plant, say 120 kW, could face penalties of up to $90,000 for three years worth of cogen gas.

So, cooperation was instituted with the cogenerators instrumenting their plants to provide the annual report to the utility on the efficacy of their systems. Newly designed cogeneration packages incorporated instrumentation built in to the package to record kilowatt output, fuel input, and thermal output, including dump-radiator output, so that the FERC efficiency could be calculated instantaneously and reporting to the utility made easy.

A typical example of this method of control and optimization is when a cogeneration plant is placed into service at a municipal swimming pool or a health and fitness club that heats their pool and spa year round. During the winter months and most of the spring and fall, the cogenerator heat is used efficiently to heat the pool water with little or no heat being "dumped.." But, when summer comes and the pool water escalates to 90 degrees from Mother Nature, the heat from the cogenerator must be redirected to the dump radiator.

It is at this very time of year that the utilities value their product most dearly by imposing higher energy charges per kilowatt as well as exorbitant demand charges. Energy charges may go from 4 cents a kilowatt in the winter to over 7 cents a kilowatt in the summer. Demand charges will go from $4.80 per kilowatt to over $25 per kilowatt of demand. The period of this high demand usually runs from May 1 thru September 30 and is bracketed from 11:00 AM to 6:00 PM daily except weekends. The utility monitors the demand meter so that if any demand increase occurs for 15 minutes during this period the full demand charge is made for the entire month. For example, if a cogenerator is producing 120 kW of electricity and the remainder of the facilities needs are 80 kW, then the facility only gets billed for the 80 kW. At $25 per kW, his monthly bill for demand would be $2,000.

However, if the cogenerator should go down for any 15-minute period during this high-demand window, then the facility gets billed not only for the 80 kW he had been getting from the utility, but also the 120 kW the utility must now send him due to his cogenerator being down, even if were only down for 20 minutes! His demand bill for that month would leap from $2,000 to $5,000, an increase of $3,000 in one month alone!

Now you see the value of the dump radiator. Should a system be getting near a default situation with regards to FERC efficiency, it would be far better to shut the system down at night or during the late fall or winter months than to have it down during periods of peak demand. The same would be true for routine maintenance such as oil and filter changes. This service should be done only at night or early morning during those high summer demand charge months.

Hence, opportunity led to greed, which in turn led to regulation. Learning to operate in a regulated environment, however, created an industry that is compatible with both the client's and the power industry's needs.

Chapter 3

Regulatory Actions

As seen from the comments made in the previous chapter, regulatory influences are felt in the field of cogeneration. Since most cogeneration facilities tie in with the local utility for standby power and, in many cases throughout the United States, receive a price break on the cost of natural gas to run the cogenerator, the government expresses their need to protect the consumer, the utility and the cogeneration facility with regulations.

The Public Utility Regulatory Policies Act of 1978 (PURPA) is a legacy of the energy policy of the late 1970's which attempted to substitute the government's fuel-choice and energy conservation judgments for those of the marketplace. The dramatic run up in oil prices by over 230% between 1970 and 1980, was viewed as a threat to both national security and economic stability. (See Table 3-1). As a result, the government instituted a number of conservation measures, the most prominent of which was PURPA.

Table 3-1

Fuel	1970	1975	1980	1985	1990
Crude oil	155.7	268.7	519.1	439.9	304.8
Natural Gas	43.8	81.7	202.0	239.1	136.5
Anthracite coal	138.6	303.9	259.3	216.3	154.0

Source: Statistical Abstract of the United States 1995, p. 596.

Enacted in 1978, PURPA was intended to conserve fossil fuels by stimulating the production of electricity from renewable sources and increasing the efficiency of electricity use. This objective was accomplished by a policy that favored, and continues to favor, renewable resource generation and cogeneration.

While PURPA has helped to stimulate the development of a nonutility generation sector, it has saddled some utilities with substantial contractual obligations for electricity supply that are not cost effective, even in today's market, and clearly will not be viable in a competitive market place. These noncompetitive PURPA contracts constitute a significant percentage of utilities' stranded costs, and are an impediment to the development of competitive electricity markets.

PURPA requires utilities to purchase power from "qualifying facilities" (QFs) at the utility's avoided cost of producing power. QFs consist of small-power producers (SPPs) using renewable resources and cogenerators. Partially as a result of PURPA, nonutility generation, which includes independent power producers (IPPs) as well as QFs, increased by 275% during the period 1978-1992. The nonutility sector now accounts for 8% of the total U.S. generating capacity.

PURPA is an excellent example of a regulatory program that, while sounding reasonable on its face, has led to serious adverse consequences. In an unregulated market, a cost-minimizing utility would purchase outside power when doing so is cheaper that its own generation. There would, therefore, be no need for PURPA-style regulation. In a regulated market, utilities may have strategic reasons not to purchase outside power (even if it is less expensive) since doing so may constitute a threat to their monopoly position. Therefore, requiring a utility to purchase power from QFs at the utility's avoided cost sounds like a policy that would simply stimulate an efficient market and would lower prices to consumers. This has not, however, been how the statute has been implemented.

Under PURPA, states have promoted QF electricity by adopting methodologies that yielded high avoided-cost estimates and required utilities to purchase QF power under unfavorable conditions. As an example of this the California Public Utilities Commission (CPUC) set up a series of power-purchase contracts that defined avoided cost as that which the utility sells to its customers. So, in the late 1980's and early 1990's, the three major California utilities were required to enter into contracts with cogenerators to buy back their excess power at prices of $.08 - $.10 per kilowatt. These Standard Offers, as the contracts were called by the utilities, caused a rampage of electrical production that was sold to the utilities under the guise of cogeneration. A plant may be set up to produce electricity and thermal energy in the form of hot water or steam and simply blow off much of the thermal energy in order to enjoy the high rates of electricity sales back to the utility while buying cheap cogeneration gas, often from the same utility as in the case of San Diego Gas and Electric and Pacific Gas and Electric, two of the largest public utilities in California.

These Standard Offer contracts were instituted when the price of fossil fuels was at or near their peak and contracts implemented at those times did not have renegotiation rights when the price of fuel went down, as it had in the late 1980's and 1990's. State utility commissions did recognize these disparities eventually and removed this onus from the utility's requirements to enter into long-term contracts at prevailing power-selling prices. They allowed utilities to define what "avoided cost" meant in the form of what it cost the utility to generate their own power, and it was this figure that would be used in buy-back Standard Offers. Of course, innovative bookkeepers within the utilities began to show "avoided costs" that were favorable to them and not necessarily the true picture. In this country the pendulum has a big swing. Now, it is not very cost efficient to sell power back to a utility that is only offering $.02 -$.03 per kilowatt.

During the high Standard Offer contract period, and there are still contracts in force that are paying those eight to ten cents per kilowatt, the utilities tried to force the power producer to prove that they were meeting the FERC efficiency guidelines of 42.5% overall efficiency. Squabbles ensued between the producer, who did not have the instrumentation to measure thermal output, and the utility as to who should monitor the FERC efficiency guidelines. Some utilities threatened that if it were shown that a producer had violated the FERC efficiency, the cogeneration gas price break that they had been enjoying would not only be retracted, but a penalty of the past three year's price break would be imposed.

For large cogenerators, 300 kW to 2.0 MW, this could mean coughing up hundreds of thousand of dollars in penalties. Many cogenerators compromised with the utility and promised to sell back only during peak usage periods when utilities had to put on their least efficient units to meet demand. Generally, peace is prevailing between cogenerator and utility at this time with the advent of the deregulation of electricity now looming as the solution to the issues of buying electricity from the low cost producer.

As a footnote to these issues, it may be enlightening to show the cost of operation of a small-scale cogeneration facility located in an area that allows gas to be purchased at the utility rate. To produce 120 kW of electricity, 10.7 therms of gas is consumed per hour. Gas costs, under the most favorable contract price, is about $.30 per therm. On gas usage alone the price to fuel the cogenerator is $3.21. Therefore, for 120 kW produced the cost per kW for fuel gas is $.0268/kW. Maintenance costs are often stated as a cost/kW and will run about $.013/kW. These two costs alone equal $.0399 or almost 4 cents per kW, and none of the investment price of the facility is yet included. You can see why an offer to buy electricity, excess or not, at 2 to 3 cents/kW is counter-productive to operating a cogeneration facility solely to sell power back to the utility.

The deregulation of electricity is now in the news. California has taken the lead in this by implementing a law, which went into effect in March of 1998, to allow any user of electricity to buy their power from anyone. This, of course, includes the utility from which they have been buying their power. Three main factors influence the cost of electricity: A) Generation, B) Transmission, and C) Distribution. The utility, by the way, will still be the transmitter and distributor of the electricity via their existing transmission and power lines strung over the countryside and the streets of the community. So it is basically the generation of electricity that will cause costs to vary from the present overall cost of electricity. Any savings to the consumer will have to come from this source and if the "avoided cost" of electricity production is 2 to 3 cents, as the utilities are claiming, then the overall savings of electricity supplied by a third party will be a percentage of those cents. That is not to say that the consumer's savings will be minuscule. Some power producers, existing utilities and others, will have an excess of power or will be able to generate at lower costs than other producers. They will pass these cost savings on to the consumer. Competition will enter this heretofore oligarchy, consumers will have choices, and inevitably savings will be realized by the consumer.

The California utilities have exacted their pound of flesh from the regulators however. Since they were previously granted rate approval based on their invested capital, this will no longer be the case. The utilities will exact a stranded cost from consumers who choose to leave their system and buy power from a third party. This "stranded cost" may be as high as 3 cents per kilowatt, but whatever it is it will only last until the year 2002. At that time, the CPUC believes the utilities will have recouped their investment of existing generation plants, and the market will be free of all encumbrances. Until the next round of regulation, at least!

And, the next round of regulation has occurred. It regulated deregulation. Since California was the first state to deregulate its utilities emphasis must be placed on that state's activities since deregulation. Other states have watched California closely in order not to emulate the situation that has occurred there.

It behooves a close look at the history of what has happened in California. In 1996 deregulation was passed by the state legislature. This allowed competition of power supply from outside California. The three major public utilities were required to choose the areas in which they wanted to operate, i.e. generation-transmission-distribution, but they could not continue to operate in all three areas. All chose to divest themselves of the generating sector by selling the existing generating stations with the exception of the nuclear facilities.

Under deregulation, consumers were given a 10% cut in electricity prices and a rate freeze until March 2002, or until utilities recovered all their past investments, whichever happened first. The utilities were allowed to issue $7 billion in bonds to pay for the 10% rate cut and charge customers over 10 years to repay the bonds. In effect, the 10% rate cut amounted to about 3%. It's important to note here that SDG&E, the utility company that services Southern California mainly in San Diego and Imperial counties with a small slice of Orange county—a relatively small area encompassing about 3.5 million people, was the first to sell its generating stations and did so at a price that paid for their past investment in those plants. Hence, they were the first utility in the country to become fully deregulated and the rate freeze could be suspended. The cost to consumers for electricity immediately increased which shocked everybody since deregulation was to be the road to lower electricity prices.

Rates in San Diego rose three-fold. What was a $72 typical monthly bill for residential consumers along the coastal area all of a sudden became $240 during the summer of 2000. Inland city residential consumers saw their bills go to over $500 per month because of their dependence on air conditioning. In the rest of California where PG&E and Southern California Edison supplied electricity to over 25 million people, the rate freeze caused both of those utilities to suffer financial hardship. By January 2001, rolling blackouts were initiated and the state was spending $40 million each day to buy power from outside the state or from the few municipalities in California that had power to spare such as LADWP.

Part of the cause for this crisis lies in the way electricity deregulation was structured in California. Prior to deregulation, those three large, privately owned utilities delivered 80% of the electricity consumed in the state. The vertically integrated utility organizations managed all aspects of the power industry, Although the utilities operated in a monopolistic venue, they did have checks and balances imposed upon them via the Public Utility Commission by scrutinizing every rate increase application. The utilities applications for rate increases were usually modest and were dictated by a reasonable rate of return on their invested assets. Electricity prices had actually gone down in San Diego Gas & Electric's market during the early 1990's. At one time SDG&E's prices were among the highest in the nation but were consistently lowered before deregulation. So, since deregulation was supposed to reduce rates, why did rates all of a sudden escalate in San Diego, and cause the other utilities to face bankruptcy?

The California deregulation bill resulted in a market structure that was very different. Under deregulation, the monopoly was broken, and separate entities were made responsible for generation, transmission and distribution. An Independent System Operator (ISO) was formed to balance supply and demand on the transmission lines. It also operates an ancillary services market to provide operating reserve in the power grid to maintain reliability. A Power Exchange (PX) was set up to determine market clearing price in the Day Ahead Market, Hour Ahead Market, and the Real Time Market.

Many questions arose after the disaster that hit the California power market:

1. Is this market system complete?
2. Does it provide all the checks and balances that were found in the system before deregulation?
3. Will the market provide for capacity needed in the future?
4. Does the structure provide for expansion of transmission capacity?
5. Does it provide for social benefits and adequate public safety?

Most of the answers are negative. Analysis of the events of the summer of 2000 and succeeding months points out several problems in the market and the inadequacy of attempted solutions. For example, during the summer of 2000, the ISO attempted to control wholesale prices by lowering the price cap. However, analysis of the market in 2000 indicated that lower price caps did not reflect the actual cost of electricity on the open market. Supply cost and scarcity outweighed the controls and pushed the cost higher.

Underscheduling also contributed to the debacle. The market need was to be balanced by real-time scheduling. It was designed to have most of the electricity (greater than 95%) traded ahead of time and leave only a small amount for the ISO to buy out of the real time market. However, during the summer of 2000, up to 16,000 MW of underscheduling occurred, requiring the ISO to purchase the deficit outside the PX market.

Of course, the real problem was supply vs. demand. In the late 1980's and early 1990's it was supposed that electric demand would abate due to more efficient appliances, conservation, utilization of other on site power sources such as wind, hydro, photo voltaics and cogeneration. No one saw the immense demand generated by the new information age: Cell phones, internet services, pagers, computers, and all the other electronic devices that by themselves only sip power, but the infrastructure to support that sipping requires huge amounts of power. The electric power industry did not anticipate this growth and did not allow for it in the planning and building of new power plants. By the time this influx of new power demand was upon California, it was too late to build more supply. Conventional supply of power, i.e. fossil fueled steam driven power plants take up to 10 years to plan and build. The new source of power—gas turbines—had lead times in the 4- to 6-year range. And, since power generation was now more or less in the hands of nonutility companies who would operate power plants for profit, was the profit going to be there if price caps were reinstituted, as they have been.

What this amounts to is the learning that can be gained from California deregulation:

- Is electricity suitable for a competitive market?
- Electricity is a necessity and provides essential services.
- It cannot be easily substituted with other energy sources at the consumer level.
- There is a long lead time to build new power plants.
- There is a high entry cost and risk for new competitors.
- Did it meet the expectations of lower electric costs for everyone?
- Does it encourage innovation?
- Does it encourage capital investment?

Since SDG&E's experience with deregulation is the most documented it is interesting to note what has occurred in that market area. As of mid 2002 SDG&E's electric rates are the highest in the state and among the highest in the country. That will be further exacerbated when a new city fee of 4% is imposed to pay for new underground power lines. The company states that the reason for these high rates is due to the crisis brought on by power deregulation. When the aforementioned increases in residential rates occurred in 2000, the clamor that ensued resulted in a lowering or capping of residential rates to 6.5 cents/kWh.

This was to have been a "loan" to the consumer, now being paid back by higher overall electric rates. SDG&E's bills to the consumer are staged with a baseline rate of 6.871 cents/kWh for the first 337 kWh of consumption, with the balance charged at 9.362 cents/kWh. A consumer using 775 kWh of electricity per month has a base charge of $64.12 or 8.27 cents/kWh. From there factors are used to charge even more using a formula of 130% of baseline (337 × 1.3 = 438) at 6.5 cents/kWh; 131-200% of baseline at 7.425 cents/kWh; and 201-300% of baseline at 8.332 cents/kWh. The total cost to the consumer therefore, for 775 kWh of power is $118.54 or 15.3 cents/kWh—a far cry from the cap of 6.5 cents/kWh that was imposed after the hue and cry of 2000.

Consumer advocates say that SDG&E customers are paying 13% more than customers of Southern Cal Edison (SCE) and 27% more than clients in the PG&E service area. Of course, neither of those two utilities are fully deregulated as of 2002 although they have been allowed significant rate increases by the PUC in order to stave off total bankruptcy. PG&E has claimed Chapter 11 bankruptcy and SCE was close to doing the same. SDG&E is buying about 50% of its electricity from the state compared to 33% for SCE and PG&E. Those state supplied kilowatts are burdened by long term contracts the state signed with outside suppliers to ward off rolling and actual blackouts which are priced at $80 per megawatt-hour vs. the spot market price today of $30 per megawatt-hour. The state justifies this anomaly in price by saying "it kept the lights burning"!

The lessons learned in California are being used by other states where deregulation is being considered, or where deregulation was imposed with much different rules. Pennsylvania is cited as a state where success in deregulation was accomplished with consumers allowed to shop for lower competitive rates, but also provided with the option of staying with their local utility at controlled rates. Officials in that state admit that their utilities are still highly regulated.

To understand where various states now stand on deregulation the

Regulatory Actions

Federal Energy Regulatory Commission has a web site, *www.fere.gov* that can be looked at. Under Informational Resources, go to Links, then to Electric Related Links and then to American Public Power Association. From there go to Legislative Regulation, then to State Restructuring and then to State Restructuring Activity Map. This map shows in color those states where restructuring of electric rates (deregulation) is active, delayed, suspended (California) or not active. As of mid 2002 17 states are active in restructuring their electric rates; 6 states have delayed restructuring; leaving 26 states where no activity is taking place on restructuring. California is listed as having suspended restructuring activity. That is one reason why it can be said that deregulation has been re-regulated!

How does this apply to the subject of cogeneration? Recognizing the need for more power and the efficiency that cogeneration supplies, the State of California has instituted a rebate program for Qualified Facilities. Under FERC, a Qualified Facility is one which meets minimum efficiency guidelines which are explained in Chapter 2, and apply only to those facilities that capture the waste heat from the prime mover to create another source of energy. Cogeneration defines that effort. The rebate program gives back to the, owner of the facility $1/watt generated or 30% of the total installed cost of the cogeneration project, whichever is lesser. It encompasses projects of 1.5 megawatts or less. The rebate includes not only the equipment and installation, but the sales tax, engineering, freight, interconnection fees and any cost that was expended to implement the cogeneration project. That program started in 2001 and will continue to 2004 and appears to be successful in seeing cogeneration projects being implemented across the state.

Chapter 4

Uses of Cogeneration

Cogeneration can be applied anywhere a facility has need of two or more energy uses. Energy uses are described as electricity, hot water, steam, chilled water, space heating, chemical bath heating, air conditioning and just about any other need that requires energy input.

The most typical use is when a facility needs electricity and hot water. Obviously, electricity is universal in its use, and rarely would we find a cogeneration system in operation that would not have electricity as one of its energy products. Hot-water applications are found everywhere, both in commercial and industrial applications. Residential use is also an area where cogeneration can be successfully applied if the user is large enough or if the technology to provide suitable cogeneration is available.

The typical water heating examples are as follows:

A) Hotels: Guest room water for bathing and showering; laundry service; kitchen service for dish washing; swimming pool heating; spa heating.

B) Restaurants: kitchen service for dishwashing, lavatory hot water.

C) Hospitals: patient room-bathing and showering, therapeutic pools, spas; swimming pools, kitchen service, laundry service.

D) Health and fitness facilities: swimming-pool heating, spa heating, showers and lavatory service.

E) Municipalities: swimming-pool heating, spa heating, lavatory and shower service.

F) Recreational-pool facilities: swimming-pool heating, water-slide areas, tubing and wave-riding water heating, shower and lavatory service.

G) Nursing homes and care facilities: patient showering and bathing, therapeutic pools, spas, kitchen Service, laundry service.

H) Coin-operated laundries: hot water for washing clothes.

I) Commercial laundries: hot water for washing clothes, uniforms, sheets, etc.

J) Metal-plating factories: hot chemical baths

K) Food-processing plants: hot water for cooking, cleaning with hot water, lavatory service.

L) Residential: swimming-pool heating; spa heating; lavatory water for showering and bathing, kitchen and laundry service.

As can be seen, any facility that has a need for hot water is a potential user of the benefits of cogeneration. There is another practical use for cogeneration when hot water is not needed in the facility to any great degree: cooling in the form of air conditioning or refrigeration. The hot water generated by the cogenerator can be used in making chilled water by using a technology called absorption-chilling.

The absorber chiller works on the principal of boiling a chemical solution in a vacuum with the resultant chemical vapor acting as a refrigerant to remove heat from water that has been used as a coolant. Once the heat is removed, the chilled water goes back to the process to cool whatever is required to be cooled, picking up heat, and returning to the absorber chiller to be chilled again. Typically, an air-conditioning system using natural gas in a burner is used to boil the chemical solution so that the refrigerant vapor is released and used to chill water. That chilled water is blown over a heat exchanger to cool the air in the room. The chilled water picks up heat from the room, that water is returned to the absorber chiller slightly warmer than when it left to be cooled again, and the cycle repeated. Instead of using natural gas in a burner to provide the heat to boil the chemical refrigerant, hot water may be used as the heat source. Lithium bromide is often used as the chemical refrigerant.

Consequently, a facility that uses electricity but little or no hot water can utilize the benefits of cogeneration via the production of electricity and hot water to be used in an absorber-chiller type air-conditioning unit.

Chapter 17 discusses absorber-chiller technology in more depth showing the typical performance of these units when fueled by the hot water output of a cogeneration system.

Steam is used in many facilities to provide space heating, in-process systems, sterilization of instruments, cooking and many other applications. Cogenerators can be designed to use the waste heat to go directly to steam, usually low-pressure steam. If you've ever seen an automobile radiator gush steam, it is from overheating the coolant. This is an indication that the coolant was not removing heat fast enough from the car's engine, so the coolant started to boil and create steam.

Often, it may be just as practical to use the cogenerator's hot-water system as a means to preheat the boiler feed water that is pumped into the boiler to make steam. A Btu is defined as the amount of heat required to raise one pound of water one degree Fahrenheit. A boiler that produces 10,000 pounds per hour of steam usually raises the temperature of the incoming water to the boiling point and then adds additional heat for the steam pressure desired. Steam systems can be either once through, meaning the steam is released to atmosphere and lost; or as a return system where most, if not all, of the steam is returned to the boiler as condensate. In those cases, the temperature of the condensate is anywhere from 140°F To 180°F and must be raised to the boiling point and beyond. If the temperature can be raised 30 to 40 degrees by cogeneration, that is less fuel required by the boiler to produce steam. Therefore, a 10,000-pound-per-hour boiler will use 400,000 Btu of energy just to raise the condensate from 150 to 190 degrees Fahrenheit. A cogenerator producing 100 kW of electricity can also produce that 400,000 Btu of thermal energy as well.

Therefore, whether a facility's needs are electricity and hot water or steam or air conditioning, cogeneration can make a significant difference in the amount of money spent on utilities within that facility.

Chapter 5

Applying Cogeneration To a Facility

While a facility that uses two or more of the forms of energy discussed in the previous chapter may be a candidate for cogeneration, only an in-depth analysis will determine if cogeneration will grant a reasonable return on the investment. Gathering the data, therefore, is a prerequisite to making any offering to that facility.

Using a water-heating facility as an example of the type of data needed, a review of the facility's history of energy usage is required. Usually a year's history is all that is needed as within that year all the seasonal changes will be identified. The provider of the facility's electricity is the first source of history. Each month the facility receives an electrical bill that defines the amount of energy used and the cost for that energy. Today, however, most utilities are metering their commercial and industrial customers on a time-of-use basis. They do this for two reasons. One is that they will have seasonal changes in their rates and, even within seasons, they will have periods of peak energy costs and off-peak or semi-peak energy costs. It is safe to say that in today's modern world air conditioning dictates the peaks of electricity costs. You can bet that the highest cost of electricity will be at noon in August in most parts of the United States.

The electric bill will break down the kilowatts of electricity used in any given month as well as the demand cost for the kilowatts that were used instantaneously in a given hour or partial hour. The first cost is usually referred to as Energy Cost; while the second is called Demand Cost. The utility figures that it must have available enough kilowatt capacity to serve the instantaneous needs of its customers. If a customer routinely stops and starts large motors in the course of their business day, their average energy usage will be determined by how long that motor was on versus how long that motor was off. If a 100-horsepower motor uses 74.5 kilowatts of electricity in an hour and is on for four hours and off for four hours, the utility will bill that customer for 74.5 kilowatts times four hours or 298 kilowatts that day. But the utility does not know what four hours the facility will have that motor on. So, it must have in reserve enough power to provide electricity whenever that facility decides to turn on that motor. The facility *demands* the power when it wants it and the utility will impose a *Demand Charge* for having enough reserve to satisfy that customer. The facility's electric bill will not only reflect that 100-hp-motor's energy charge, but also show a demand charge of 74.5 kW for that month.

One southern California utility will charge that customer $.0735 per kilowatt for the energy used in running that 100-hp motor at noon in August, and $26.50 per kW of demand for that month. Seventy-four and a half kW times $26.50 = $1974.25, which will be billed to that facility over and above the energy costs.

Residential utility bills only show total kilowatts used and the cost for purchasing same. However, in most residential bills the cost of energy is twice that billed to a commercial or industrial customer. But, the commercial and or the industrial customer also pays that Demand Charge.

Utilities figure that the average residence will not be turning on and off large motors during the day. Most residential air-conditioning units use a 5- to 10-horsepower motor at best, which would be the largest single electricity user in the household.

Therefore, the first part of the analysis is to break out the energy costs from the demand costs and list them by month along with the energy and demand used. Then, a load profile must be considered. A load profile is an hour by hour picture of when the electricity is used within the facility. If the facility opens at 6:00 a.m. and closes at 10:00 p.m., it is reasonable to assume most of its electricity is consumed within those hours. But, if air-conditioning units are left on all night or swimming-pool circulating pumps run all night and day, plus security lighting and whatever else may still be in use, the electric load should be looked at for the entire 24-hour day. The daytime load may support a 120-kilowatt cogeneration unit, but the night time load may fall to only 20 kilowatts of demand. Is it worth running the large cogenerator during the night only displacing 20 kilowatts of electricity and selling or giving back to the utility the other 100 kilowatts? Probably not. Therefore, prudence suggests shutting down the cogenerator when the facility closes and restarting it when the facility opens.

However, if the facility's night time load is 60-kW demand, then that same prudent analyzer might offer two 60-kW cogenerators, both to run during the day and only one at night. Analysis will determine the optimum offering to the customer.

Once the electric bills have been analyzed and the facility's usage pattern developed, attention is then turned to the gas usage, or more appropriately, the water-heating usage. If the facility is heating a swimming pool, a spa, and shower and laundry water via water heaters running on natural gas, then the gas usage is at least partially attributed to those functions. But is there any other gas usage taking place within the facility. Space heating? Cooking? Laundry drying? These functions must be taken into account so it can be determined how much of the total gas consumed is for water heating versus other uses that will not be displaced by cogeneration.

Comparing the summer gas usage with winter may be all that's needed to see what significant space heating is taking place. But then again, the pool requires less heat in the summer than in the winter to heat it to a comfortable level, so the total gas consumed in August is going to be far less than that consumed in February. Looking at the space-heater's specifications and computing its gas consumption is probably the safest way to determine what gas is being consumed. Doing the same with the pool's heater will also tell how much gas it consumes. The utility's log of degree days will then allow the analyzer to determine how many hours in a given month the space heater operated as well as the pool heater. For other gas-using devices, such as the cooking area and the laundry dryer, their specifications and hours of average daily usage will allow sufficient estimates of their gas use.

Once that analysis has been accomplished, it is easy to determine how much gas is being used to heat water that can be displaced by cogeneration. Take the total gas consumed in the past year and multiply it by the percentage of gas used for water heating. That number is what the cogenerator will attempt to offset.

We now have the total electric consumption in energy and demand and the total gas consumed in heating water for the facility. The next step is to apply a cogenerator that will displace as much of those utilities as possible without infringing upon the minimum FERC efficiency of 42.5%. This is important because most facilities will use significantly more electricity than gas, and the first tendency is to offset as much of that electricity as possible. But, if the generated thermal energy is significantly greater than the facility is using, thermally, then an upside down condition may arise where the FERC efficiency criteria is not met. This is important as most utilities will require an audit of the total energy generated versus the fuel used in the cogenerator, and with a dump radiator on site, the utility will require the heat exhausted via the dump radiator to be included in the FERC calculations. Examples of these conditions will be found in Chapter 6, Sizing the Cogenerator.

Chapter 6

Sizing the Cogenerator

By using the data gathered from the utility bills and the analyzing of that data, a cogenerator can be selected that meets the optimum needs of the facility. One of those optimum needs is not to overproduce either electricity or thermal energy. The other need is to see that the economics are suitable to the customer. Whatever altruistic benefits derive from cogeneration, i.e. environmental and fuel conservation, customers will not be attracted to spending their money unless it shows a fair return on their investment.

The computer is a valuable tool in making this selection. Variables in energy production and hourly operation, are easily handled by a computer with the results readily viewable. The following analysis is from a project where the client's actual electrical and gas usage as well as the cogenerator's specifications have been entered to present a complete picture.

Table 6-1 shows that this customer has consumed 199,264 therms during the year, which cost $122,152; Table 6-2 shows 4,390,560 kilowatts used at a cost of $307,339 for the energy; and Table 6-3 shows a demand usage of 10,284 kW, which cost $102,840.

Table 6-4 depicts how this gathered information is put into the computer and is analyzed.

The computer calculates the cost per therm of the gas; the cost per kilowatt of the electrical energy and the cost per kW of the demand, Lines A-1, A-2 and A-3. The thermal-load factor (amount of gas the cogenerator will attempt to displace) and the efficiency of the heaters used to heat the water are put into the computer, Lines A-4 and A-5.

Line A-6 relates to the hours in which the facility is open on an annual basis. For instance, a hotel is a "24-hours-a-day, 7-days-a-week" business. Even at night, the load factors are quite high in both electrical and thermal load. However, it is usual to select a factor of 90% of the actual hours in a year. So, of the 8760 actual hours in one year it is prudent to use only 7800 hours of cogenerator run time. This allows for

Table 6-1

MONTH	GAS IN THERMS	DOLLARS
JANUARY	23,456	$14,379
FEBRUARY	22,867	14,017
MARCH	24,398	14,956
APRIL	18,985	11,638
MAY	15,763	9,663
JUNE	12,457	7,636
JULY	8,956	5,490
AUGUST	4,765	2,921
SEPTEMBER	6,789	4,162
OCTOBER	14,374	8,811
NOVEMBER	22,980	14,087
DECEMBER	23,474	14,392
TOTAL	199,264	$122,152

Table 6-2. Electric Energy Usage

MONTH	ELECTRIC USAGE KILOWATTS	DOLLARS
JANUARY	294,500	$20,615
FEBRUARY	260,890	18,262
MARCH	306,000	21,420
APRIL	330,980	23,169
MAY	356,700	24,969
JUNE	389,760	27,283
JULY	458,380	32,086
AUGUST	484,680	33,928
SEPTEMBER	513,470	35,943
OCTOBER	368,590	25,801
NOVEMBER	332,450	23,272
DECEMBER	294,160	20,591
TOTALS	4,390,560	$307,339

Sizing the Cogenerator

Table 6-3. Electric Demand Usage

MONTH	DEMAND USAGE KILOWATTS	DOLLARS
JANUARY	480	$2,400
FEBRUARY	430	2,150
MARCH	490	2,450
APRIL	520	2,600
MAY	760	9,120
JUNE	950	15,200
JULY	1,120	17,920
AUGUST	1,340	21,440
SEPTEMBER	1,244	14,810
OCTOBER	1,080	5,400
NOVEMBER	980	4,900
DECEMBER	890	4,450
TOTALS	10,284	$102,840

scheduled and unscheduled maintenance of the cogeneration system.

The computer has the operating data for the manufacturer's cogeneration sizes in its memory and suggests the size of the system for both the electrical load as well as the thermal load. In the case of our example it is suggesting multiple 120 kW units to meet the electrical load, Lines A-7 and A-8.

Lines A-9 through A-14 represent the rating of the cogenerator(s) in kilowatt production; gas used to run the cogenerators (thermal input); useful thermal energy produced (thermal output); and then the overall efficiency of the system if all the energy is used on site is calculated. In this case it is 88%, Line A-12.

Line A-13 is the computer's calculation of the number of hours needed to satisfy the thermal needs of the facility. This case says over 11,000 hours are needed, which means that the selection of two (2) 120-kW units is conservative. The system will not overproduce thermal energy when running 7800 hours per year. Nor will it overproduce electrical energy either. Of the 4,390,560 kW used, the cogeneration system will produce 43% or 1,872,000 kW.

Table 6-4

Line		
A-1	Is this Cost/Therm OK?	$0.613
A-2	Is this Cost/kWh OK?	$0.070
A-3	is this Cost/Demand OK?	$10.00
	Is Site Time of Use Metered? yes	
	Will Run Time be in all periods? yes	
	Demand Standby Charge/kW	0.000
	Utility Sell Back Rate	0.000
A-4	Thermal Load Factor	80.00
A-5	Present Efficiency	80.00
	Closed Thermal Load Percentage	0 %
	Closed Electric Load Percentage	0 %
	A/C kWh Displaced per year	0
	A/C Demand Displaced per year	0
	Useful Thermal Increase per year	0
A-6	Maximum hours open per year	7800
	Average therms/open hour	16.35
	Average kWh/open hour	562.89
	Average peak demand/month	857.00
A-7	Thermal Load suggests a 120	
A-8	Electric Load suggests a .... 120's	
A-9	kW Output Setting	240.00
A-10	Cogen Thermal Input	22.00
A-11	Cogen Thermal Output	11.22
A-12	Total Efficiency	0.88
A-13	Run Hrs to Meet Thermal Need	11366
A-14	Estimated Run Hours	7800
	RUN HOURS EQUAL OR ARE LESS THAN HOURS	
A-15	Gas Cost Eliminated is:	68.62 $
A-16	Current kWh Consumed	4390560
A-17	Generated kWh	1872000
	NOTE—GENERATED KWH DOES NOT EXCEED 90% OF	
A-18	Electric Cost Eliminated is:	43 %
A-19	FERC Percentage (42.5% min.)	62.75
A-20	Cogen Gas Rate ($/therm)	0.290

Sizing the Cogenerator

Table 6-4 (*Continued*)

Line

	TAXABLE MATERIAL COSTS:
COGENERATION UNIT:	$178,000
HEAT EXCHANGERS:	$8,000
DUMP RADIATOR:	$3,800
PUMPS (2):	$1,600
STORAGE TANK:	$0
PIPE, VALVES & FITTINGS:	$9,000
EXHAUST PIPING:	$300
ELECTRICAL MATERIALS:	$9,000
ELECTRIC METER:	$300
GAS METER:	$0
WATER FLOW METER:	$0
THERMAL & METER SENSORS	$0
MISC.	$1,000
CATALYTIC CONVERTER	$12,000
SUBTOTAL TAXABLE MATERIAL COST:	**$223,000**
STATE TAX RATE.,	7.75 %
APPLICABLE STATE TAX:	$17,282.50
ADDITIONAL TAX RATE:	0.00 %
APPLICABLE ADDITIONAL TAX:	$0.00
TOTAL TAXABLE AMOUNT:	$17,282.50

	ADDITIONAL INSTALLATION C
ENGINEERING:	$8,000
LABOR:	$10,000
PERMITS:	$4,500
UTILITY/INTERCONNECTION:	$35,000
SHIPPING:	$1,500
RIGGING:	$1,500
SALES COMMISSION.*	$0
G & A COSTS	$130,000
SUBTOTAL ADDITIONAL COSTS:	**$190,500**

Table 6-4 (*Continued*)

Line		
	TOTAL COSTS (LESS TAXES):	$413,500
	UTILITY REBATE/CREDIT AMOUNT:	$0
	FIRST YEAR MAINTENANCE COST:	$22,000
A-21	TOTAL PROJECT COST:	$452,783
	PROJECTED PAYBACK (YEARS):	2.34
	LEASE PERCENTAGE RATE:	10.00 %
	LEASE TERM IN MONTHS:	60
	LEASE PAYMENT:	$9,620
	DEPRECIATION TAX BRACKET:	30 %
	Monthly savings WITHOUT Dep.:	$14,041
	Over/Under lease payment?	$4,421
	Monthly Savings WITH Dep:	$16,305
	Over/Under lease payment?	$6,685

Line A-19 shows a calculation of the FERC efficiency at 62.75%, well over the minimum requirement of 42.5%. It is very possible a third unit of 120 kW capacity could be added to this system with no adverse affects of overproduction or FERC efficiency. For other reasons, such as nighttime loads, the analyzer decided to offer the 240-kW system consisting of two 120-kW units.

The next analysis is the comparison of the system costs versus the savings to determine what kind of return on investment the customer might enjoy if he decides to install this system. The costs come from the manufacturer's price lists and field analysis of the engineering, plumbing, wiring, controls, overhead, etc. to determine a final proposal price. This project would carry a total turnkey price to the customer of $452,783, Line A-21. The word turnkey applies to the fact that all facets of the installation are covered in this price and no hidden or future costs will be borne by the client. The provider installs a complete system, and when finished, he "turns the key" over to the customer as the official owner.

In this case the provider has also offered a maintenance contract to the customer at a cost of $22,000 per year. This contract will cover all routine preventive-maintenance chores such as oil and filter changes, tune-ups and the like; as well as any replacement parts or *components* the system may need over the life of the maintenance contract. That includes

Sizing the Cogenerator

engine, generator, pumps, controls and any other included in the provider's contract. This is a very common maintenance contract often referred to as an *Extended Warranty and Maintenance Contract*, very similar to those offered by appliance dealers and manufacturers for one's home refrigerators, air conditioners, etc.

Tables 6-5 and 6-6 are the compilation of the input data and the analysis to show the client what he can expect in the way of energy savings and their associated costs versus the cost of the installed system and maintenance contract. Section B of the First Year Cost/Savings Calculations is a repeat of the customer's utility consumption and the thermal load that the cogenerator will displace.

Section C of this analysis is a computation of the cogenerator's savings of both thermal (gas) costs, electrical-energy costs and electrical-demand costs. The analysis shows the cogenerator offsetting 87,516 therms of gas at a savings of $67,061. The 1,872,000 kW of electricity offset will save $131,040. The demand generation of 2880 kW will save another $20,160. Note in the case of demand savings, only 70% of the total demand is taken as a savings credit. The reason for this is that as is pointed out in Chapter 2, if the cogenerator should be down for any 15-minute period during the operating month, the demand credit is lost. Most manufacturers and providers account for the fact that unscheduled downtime may occur during a peak-demand period losing the demand credit for that month. Scheduled maintenance is usually done during the off-peak billing periods for obvious reasons. Even so, the demand savings of $20,160, while important, is only about 10% of the total energy savings of $218,261.

Section D shows the gas used by the cogeneration system during its 7,800 hours of operation. Note that the cost of gas per therm is only $0.29 versus the commercial rate of $0.613 per therm shown in Section B and on Line A-1. This is part of the PURPA agreement for qualified cogeneration facilities that provides the same gas cost to a cogenerator as that of the utility. In other words, the utility pays $0.29 per therm for its gas, and so shall the *Qualified* cogeneration *Facility*.

Section E summarizes the First Year Cost/Savings. The kWh savings are $131,040, Line E-1; the Demand Savings are $20,160, Line E-2; the gas savings are $67,061, Line E-3; for a total gross savings of $218,261. But, from these gross savings are deducted the cost of operation: fuel cost and maintenance costs for the cogenerators. Cogen fuel cost is $49,764, Line E-4 as well as D-5.

Table 6-5
FIRST YEAR COST/SAVINGS CALCULATIONS

CLIENT NAME:	BUSINESS CLUB
SITE ADDRESS:	1000 Main Street
	Los Angeles, CA 90071
PROPOSED EQUIPMENT:	Two (2) Model ISI-120I
PREPARED:	02-Nov-98

—B—CURRENT GAS CONSUMPTION:

Annual Gas Expense	$122,152	
Current Gas Rate	$0.613	$/therm
Annual Gas Usage	199,264	therms/yr
Avail. Hot Water Load	80	%
Hot Water Heater Input	159,411	therms/yr
Cost of Hot Water Gas	$97,722	

—C—COGENERATION GAS SAVINGS:

Heater Efficiency	80	%
Heater Offset Required	127,529	therms/yr
Added Thermal Load	0	therms/yr
COGEN Thermal Output	11.220	therms/hr
Thermal Run Hrs Rqd.	11,366	hours/yr
Max. Avail. Run Hours	7,800	hours/yr

—B—CURRENT ELECTRIC CONSUMPTION:

Annual Elec Expense	$307,339
Current kWh Rate	$0.0700
Annual kWh Used	4,390,560
Annual Demand Expense	$102,840
Average Demand Rate	$10.00
Annual Demand kW Used	10284

—C—COGENERATION KILOWATT PRODUCTION:

Annual Run Hours	7,800
Elec Output (kW)	240.0
Annual kWh Generated	1,872,000
Current Rate/kWh	$0.0700
Total kWh overgenerated	0
Utility Sell Back Rate	$0-.00

Table 6-5 (Continued)

Gas Offset With Cogen	87,516	therms/yr	Value of kWh Produced	$131,04	
True Gas Offset Value	$67,061				

—D—COGENERATION OPERATIONAL DATA:

Max. Avail. Run Hours	7,800	hours/yr
COGEN Gas Input	22.000	therms/hr
Annual COGEN Gas Usage	171,600	therms/yr
Est COGEN Fuel Rate	0.290	$/therm
Est COGEN Fuel Cost	$49,764	

CREDITS & REBATES:

Displaced kWh	$0	
Displaced Demand	$0	
Project Rebate	$0	
TOTAL CREDITS/REBATES	$0	
Overall Efficiency	88.23	%
FERC Efficiency	62/75	%

—C—COGENERATION DEMAND PRODUCTION:

Annual Demand Generated	2880	
Current Demand Rate	$10.00	
Est. 70% Demand credit	$20,160	
Standby Charge/kW	$0.00	
Annual Standby Charge	$0	
Demand Credit Available	$20,160	

FIRST YEAR COST/SAVINGS SUMMARY:

kWh Savings	$131,040	—E1
Demand Savings	$20,160	—E2
Gas Cost Eliminated	$67,061	—E3
Cogen Fuel Cost	($49,764)	—E4
Annual Maint Cost	($22,000)	—E5
Credits/Rebates	$0	
GENERATED SAVINGS	$146,497	—E6

Proposal offered by ISI and is prepared using data supplied by the individual client. ISI can not be held accountable for any inaccuracy. Client should consult their own Accountant.

Annual maintenance costs are $22,000, Line E-5. The overall Generated *Net* Savings are $146,497 for the first year.

Table 6-6 shows the client what the expected savings are in the first five years of operation allowing for 5% inflation of both savings and operating expenses during those five years. One additional savings item has been added in this summary: depreciation. Since the cogeneration system will be a fixed asset of the client's property, it is allowed to be depreciated over five years and will result in a subsequent tax savings to the client. The analysis shows the client to be in the 30% tax bracket, but his accountant should have the final say in the actual depreciation savings.

The bottom line for this particular client is that for an outlay of $452,783, which includes the first year maintenance costs, he can expect savings over the five-year period of $967,322. When that number is divided by 5 years, the average annual savings is $193,464. Return on Investment is then $193,464 divided by the Investment of $452,783 to show a 42.73% ROI. This also calculates out to a simple payback of investment of 2.3 years.

This is not only a reasonable return on a clients capital investment, it is one that must be noticed and discussed very seriously within that client's management. During those discussions, certain aspects of the overall installation, permitting, regulations, stability of the maintenance provider, and financing will be discussed. The following chapters will consider those aspects.

The program used to analyze a cogeneration project is available from the author at a nominal charge. For those that are interested in obtaining the program contact the author at 760-431-0930 or via e-mail at *bkolanowski@adelphia.net.*

DETERMINING THE AVERAGE COST OF ELECTRICITY

The cost of electricity used in the aforementioned example were simply the compilation of kilowatt hours used per month and the cost of those kilowatt hours. Simple division showed that the client was paying 7 cents/kWh energy charge and $10/kW demand charge.

However, if the client were running less than 24 hours per day or if the past utility bills were not a true reflection of current rates then using the simple division of past cost of electricity divided by kilowatt consumption could taint the outcome. Therefore, it is incumbent upon any analysis to insure the correct figures are used to show the client

Table 6-6

FIVE YEAR PROJECTED COST/SAVINGS

CLIENT NAME: BUSINESS CLUB
SITE ADDRESS: 1000 Main Street
Los Angeles, CA 90071
PROPOSED EQUIPMENT: Two (2) Model ISI-120I

YEAR	ONE	TWO	THREE	FOUR	FIVE
		ANNUAL SAVINGS			
ELECTRIC	$151,200	$158,760	$166,698	$175,033	$183,784
GAS	$67,061	$70,414	$73,935	$77,631	$81,513
DEPRECIATION	$27,167	$27,167	$27,167	$27,167	$27,167
GROSS SAVINGS	$245,428	$256,341	$267,799	$279,831	$292,464
		OPERATING EXPENSES			
COGENERATION GAS	$49,764	$52,252	$54,865	$57,608	$60,488
MAINTENANCE	$0	$23,100	$24,255	$25,468	$26,741
TOTAL EXPENSE	$49,764	$75,352	$79,120	$83,076	$87,230
NET ANNUAL SAVINGS	$195,664	$180,989	$188,680	$196,755	$205,235
ACCUMULATED SAVINGS	$195,664	$376,652	$565,332	$762,087	$967,322

INSTALLATION COST (LESS REBATE)	$430,783
FIRST YEAR MAINTENANCE COST	$22,000
TOTAL PROJECT COST	$452,783
DEPRECIATION TAX BRACKET	30 %
ESTIMATED INFLATION RATE:	5 %
FIRST YEAR MONTHLY SAVINGS:	$16,305
RETURN ON INVESTMENT:	42.73 %
PAYBACK PERIOD (YEARS):	2.3

exactly what his savings will be based on the exact cost of the electricity he is presently paying his utility provider.

A method to correct this inaccuracy is to use a weighted average to determine his cost of power during all time periods and all seasons. Refer to the following tables 6-7 and 6-8. These tables show a rate structure for a California utility, Southern California Edison, that is published on their web site. It is the TOU-GS-2 General Service—Demand Metered rate structure.

Table 6-7 shows the Facilities Related Component as it applies to the Demand Charge in Summer and Winter. Note that the Summer Demand and Winter Demand charge are equal at $5.40/kW. However, for the Time Related Component, which is added to the Facilities Related Component, is different for Summer and Winter. Those demand charges are broken down into On Peak, Mid Peak and Off Peak Billing Demand. Note also that the client has the choice of Option A or Option B. The difference is that Option A has a lower On Peak Demand charge, but Table 6-8 shows that Option A also carries a higher On Peak Energy Charge. The customer has already chosen what Option he is on, so it is not necessary to compare what his optimum option is for this exercise.

Table 6-8 also shows that the summer season commences at 12:00 midnight on the first Sunday in June and continues until 12:00 midnight of the first Sunday in October of each year—approximately 4 months; while the Winter season encompasses the other 8 months of the year.

So, regardless of the hours per day the customer operates, his Demand Charges will be consistent with the utility's schedule and the Option which he is on. Assuming he is on Option A, we want to determine what his average annual Demand Charge will be for the purposes of our analysis—providing the client does operate year round. The weighted average would be calculated as follows:

Summer: (Facilities Component + Time Related Component) × # of months = ($5.40 + $7.75) × 4 summer months = **52.6 (call this a factor)**

Winter: Facilities Component × of months = $5.40 × 8 winter months = **43.2 (call this a factor)**

Combine the two factors: **52.6 + 43.2 = 95.8 divided by the entire 12 months in the year = $7.98/kW of average demand for the entire year.**

It is much easier to input this number in the computer program than trying to handle the two different numbers within the same program.

Southern California Edison
Rosemead, California

Revised Cal. PUC Sheet No. 23871-E
Cancelling Original Cal. PUC Sheet No. 20197-E

Schedule TOU-GS-2 Sheet 1
TIME-OF-USE - GENERAL SERVICE - DEMAND METERED

APPLICABILITY

Applicable to single- and three-phase general service including lighting and power, except that the customer whose monthly Maximum Demand, in the opinion of the Company, is expected to exceed 500 kW or has exceeded 500 kW for any three months during the preceding 12 months is ineligible for service under this schedule. Effective with the date of ineligibility of any customer served under this schedule, the customer's account shall be transferred to Schedule TOU-8. Further, any customer served under this schedule whose monthly Maximum Demand has registered 20 kW or below for 12 consecutive months is also eligible for service under Schedules GS-1 or TOU-GS-1. Service under this schedule is subject to meter availability. Customer must elect either Option A or Option B of the Time Related Component of Demand and the Energy Charge.

TERRITORY

Within the entire territory served.

RATES

	Per Meter Per Month	
	Summer	Winter
Customer Charge	$79.25	$79.25
Demand Charge (to be added to Customer Charge):		
Facilities Related Component:		
All kW of Billing Demand, except that the Billing Demand shall not be less than 50% of the highest Maximum Demand established in the preceding eleven months, per kW	$5.40	$5.40
Time Related Component (to be added to Facilities Related Component):		
Option A		
All kW of On-Peak Billing Demand, per kW	$7.75	N/A
Plus all kW of Mid-Peak Billing Demand, per kW	$2.45	$0.00
Plus all kW of Off-Peak Billing Demand, per kW	$0.00	$0.00
Option B		
All kW of On-Peak Billing Demand, per kW	$16.40	N/A
Plus all kW of Mid-Peak Billing Demand, per kW	$2.45	$0.00
Plus all kW of Off-Peak Billing Demand, per kW	$0.00	$0.00

(Continued)

(To be inserted by utility)	Issued by	(To be inserted by Cal. PUC)
Advice 1245-E-B	John R. Fielder	Date Filed Dec 23, 1997
Decision 97-08-056	Senior Vice President	Effective Jan 1, 1998
1C2		Resolution E-3510

Table 6-7.

Southern California Edison
Rosemead, California

Revised   Cal. PUC Sheet No. 29253-E
Cancelling Revised   Cal. PUC Sheet No. 23872-E

Schedule TOU-GS-2   Sheet 2
TIME-OF-USE - GENERAL SERVICE - DEMAND METERED

(Continued)

<u>Rates</u> (Continued)

	Per Meter Per Month	
	Summer	Winter

Energy Charge (to be added to Demand Charge):
  Option A
    All On-Peak kWh, per kWh .... $0.29601   N/A
    All Mid-Peak kWh, per kWh $0.11763 $0.12961
    All Off-Peak kWh, per kWh $0.09421 $0.09421
  Option B
    All On-Peak kWh, per kWh .... $0.21296   N/A
    All Mid-Peak kWh, per kWh $0.11763 $0.12961
    All Off-Peak kWh, per kWh $0.09421 $0.09421

The above charges used for customer billing are determined using the components shown in the Rate Components Section following the Special Conditions Section.

CARE customer rates exclude any increases to rates authorized in Commission Decision Nos. 01-05-064 and 01-01-018.

<u>SPECIAL CONDITIONS</u>

1. Time periods are defined as follows:
   - On-Peak: Noon to 6:00 p.m. summer weekdays except holidays
   - Mid-Peak: 8:00 a.m. to Noon and 6:00 p.m. to 11:00 p.m. summer weekdays except holidays
     8:00 a.m. to 9:00 p.m. winter weekdays except holidays
   - Off-Peak: All other hours.

   Holidays are New Year's Day (January 1), Washington's Birthday (third Monday in February), Memorial Day (last Monday in May), Independence Day (July 4), Labor Day (first Monday in September), Veterans Day (November 11), Thanksgiving Day (fourth Thursday in November), and Christmas (December 25).

   When any holiday listed above falls on Sunday, the following Monday will be recognized as an off-peak period. No change will be made for holidays falling on Saturday.

   The summer season shall commence at 12:00 a.m. on the first Sunday in June and continue until 12:00 a.m. of the first Sunday in October of each year. The winter season shall commence at 12:00 a.m. on the first Sunday in October of each year and continue until 12:00 a.m. of the first Sunday in June of the following year.

2. Voltage: Service will be supplied at one standard voltage.

(Continued)

(To be inserted by utility)   Issued by   (To be inserted by Cal. PUC)
Advice   1545-E   <u>John R. Fielder</u>   Date Filed   May 22, 2001
Decision   01-05-064   Senior Vice President   Effective   Jun 3, 2001
2C13   Resolution

Table 6-8.

Sizing the Cogenerator

ENERGY CHARGE WEIGHTED AVERAGE

Table 6-8 shows the Summer and Winter energy charges for both Options in On Peak, Mid Peak and Off peak rates. Note that the Winter schedule does not have an On Peak rate. The periods are further defined as follows:

On Peak: Noon to 6:00 PM summer weekdays except holidays
Mid Peak: 8:00 AM to Noon and 6:00 PM to 11:00 PM summer
 Weekdays except holidays
Off Peak: All other hours.

In order to determine exactly what energy charge the client is paying one needs to know what hours of the day, days per week and weeks per year his facility operates. Once that is known then applying the proper rates to the appropriate times is needed to determine the average cents/kWh the client is paying during his operational time. For purposes of this example we will assume the client operates 24 hours per day, 7 days per week year round.

Summer Hours, On Peak: 4 months × 30 days/month = 120 days. Divide by 7 days/week = 17.14 weeks × 5 days per week = 85.71 days. Multiply by 6 hours per day for the On Peak time period = 514.28 hours. The kWh rate for this period is $0.29601/kWh. Multiply 514.28 × $0.29601 = **152.23** (call this a factor)

Summer Hours, Mid Peak: 4 months × 30 days/month = 120 days. Divide by 7 days/week = 17.14 weeks × 5 days per week = 85.71 days. Multiply by 9 hours per day for the Mid Peak time period = 771.39 hours. Multiply by the kWh rate for this period 771.39 × $0.11763 = **90.74** (call this a factor)

Summer Hours, Off Peak: Since the 4 month period = 120 days and each day has 24 hours there are 2880 hours in the Summer period. Subtract the hours of On Peak and Mid Peak from that number to arrive at the number of hours in the Off Peak period. I.e. 2880 − 514.28 − 771.39 = 1593.91 hours × $0.09421 = **150.16** (call this a factor)

Now: Add up the Summer factors: **152.23 + 90.74 + 150.16 = 393.13 and divide by the total number of hours in the Summer period, 2880, = $0.1365/kWh.**

Winter Hours, Mid Peak: 8 months × 30 days/month = 240 days, Divide by 7 days/week = 34.29 weeks × 5 days per week = 171.43 days. Multiply by 11 hours per day for the Mid Peak Winter period = 1885.71 hours. The kWh rate for this period is $0.12961/kWh. Multiply 1885.71 × $0.12961 = **244.41** (call this a factor)

Winter Hours, Off Peak: In the 8 month period × 30 days per month there are 5,760 hours in the Winter Period. Subtract the hours of Mid Peak from that number to arrive at the number of hours in the Off Peak period. I.e. 5760 − 1885.71 = 3874.29 hours × $0.09421 = **364.99** (call this a factor)

Now: Add up the Winter factors: **244.41 + 364.99 = 609.40 and divide by the total number of hours in the Winter period, 5760, = $0.1058/kWh.**

To determine the average annual rate simply use the weighted average for each season:

$0.1365 × 4 summer months .546
$0.1058 × 8 winter months .8464
Add the two = 1.3924 and divide by the 12 months = $0.1160/kWh to arrive at the annual average rate the client pays when operating 24 hours per day, 7 days per week, year round.

This is the figure then used in the analysis by multiplying that number by the overall kilowatt consumption for the entire year to determine exactly what the client will pay for that energy consumption.

This method may be easily used when the period of time the client operates is less than 24/7. If he is mainly operating a one shift plant, 5 days per week, it is obvious his overall kWh rate is going to be much higher since the lower energy rate during off peak hours of night time and weekend days will not be used. That factor alone, often makes a single or two shift scenario a better payback than the 24/7 scenario.

The changes in utility prices imposed by deregulation, re-regulation and simply requests by utilities for rate increases makes it imperative that the client knows exactly what on site cogeneration will save him. Collecting his past 12 months utility bills will tell you his consumption, but not necessarily the cost for that consumption in the next 12 months. Know his rate structure and price the displaced kilowatts accordingly.

Chapter 7

Logistics of Installation

Installation of the cogenerator involves placing the system as near to the points of electrical and thermal distribution as possible.

The electricity produced by the cogenerator will be fed into the main electrical panel on the client's premises. This is the same electrical panel that the utility uses to distribute its electricity to the facility. Does the panel know which electricity to distribute first? Well, actually, yes, it does. The electricity coming from the local source, the cogenerator, has less resistance and will be accepted fully by the panel and distributed throughout the facility. In the majority of cases the panel will also be accepting electricity from the utility and be distributing that power as well. Once through the panel the electrons will be flowing in a commingled state to wherever power is required: lights, motors, computers, switches, etc.

On the rare occasion when the facility's electrical load can be fully handled by the power coming from the cogeneration system, the utility's power will not be needed and, therefore, will not be admitted to the distribution panel. An observer will actually see the wheel inside the utility's meter stop turning: the client is purchasing no utility supplied power at that time.

Even rarer, in a properly applied cogeneration system, the facility's demand may be less than that supplied by the cogenerator. What happens to that excess power? It is distributed to the utility and literally goes back through the wires to find a home elsewhere. If the client has no sell-back agreement with the utility, this excess power is given free to the utility. With a sell-back agreement, a separate meter is installed so that when excess power is available, it is metered, and the facility gets paid for the power it supplied to the utility. As discussed in previous chapters, that payment is the utility's "avoided cost," which is quite small and not worth selling to the utility as a reason to oversize or improperly size the cogenerator.

One curious anomaly that occurred in the early days of small-scale cogeneration was that the main import meter to the facility was not detented. Therefore, whenever excess power was supplied to the utility, this main meter would physically run backwards taking off power that had previously been supplied to the facility, not only previously supplied, but which would have been included in that month's bill. In those cases the "credit" for sell-back power was equal to the full cost of the utility supplied power and not just the avoided cost. It took an installation that literally ran that meter backwards so much that it went past the zero point for the utility to realize what was happening. From then on all their meters were "detented" meaning they could not run backwards.

Getting the cogenerated electricity to the facility's main distribution panel requires wiring and conduit of sufficient size to meet the local electrical codes. Inside the panel a circuit breaker must be installed to safeguard the panel from any overloads. The panel itself must have ample capacity to accept this "new" power. In older facilities the panel may not have sufficient capacity or room to install a suitable circuit breaker. In those cases an additional panel may be required, and this will impact on the overall cost of the cogenerator system detracting from the client's return on investment. It is therefore prudent to investigate these factors during the initial walk through of the facility when gathering the data.

While locating the cogeneration system close to the electrical distribution panel is recommended, the other side of the coin must also be looked at: the thermal distribution system. As discussed previously, the engine's coolant fluid is used as the main heat-transfer medium to the various water-heating needs of the facility. Generally, the potable-water supply goes into a room or enclosure where the main water heaters are located. These can be as simple as the water heaters one has in his home, or as complex as a steam boiler. Usually the water heaters are large, industrial type heaters that take in cold water from the city water supply and heat it for distribution to whatever the facility needs. The facility's needs, however, may be quite varied. In a hotel, for instance, the hot water may be needed at 120°F for guest rooms and other lavatory services; 140°F for laundry services and maybe 160°F for dishwashing service. The cogenerator is capable of meeting these temperatures, but the water must be tempered so as not to provide 160-degree water to the guest rooms. This is usually done via tempering valves that mix cold water with hot to provide the proper temperature for the service intended. There are various fail-safe tempering valves that are commercially available to prevent a guest in a hotel room to be subjected to a 160-degree shower.

Staying with the hotel example, other water-heating needs may be the swimming pool that likes to be kept at 80 degrees and the spa or Jacuzzi that needs to be 104 degrees to meet local health standards. Often those heaters are located near the pool and spa and not in the potable-water heater room. In this example the water-heating needs make for a logistical nightmare if not handled properly. That nightmare turns into additional cost that could ruin the cogeneration system's overall cost-to-savings ratio and squelch any hope of installing a system in that client's facility.

One rule of thumb is to locate the cogenerator nearer to the plumbing side of things rather than the electrical. It is far less expensive to run copper wiring and aluminum conduit long distances rather than copper piping with its attendant valves, fittings and controls. Plus, the wiring to provide electricity is a one-way route, from the cogenerator to the panel, whereas the plumbing is a two-way street. The water has to be heated near the cogenerator, pumped to the water heater, and returned to the cogenerator—twice as much plumbing. Since all potable-water systems require copper piping by city code, it is again best to place the cogenerator closest to the potable-water heater room. Nonpotable water, i.e. swimming-pool and spa water may be carried via plastic or PVC pipes that are glued together. If nonpotable water supplies are to be heated by the cogenerator, the plumbing to accomplish that is usually less expensive.

The objective of distributing the cogenerator's thermal energy, i.e. hot water, is to relieve the facility's existing hot-water heaters of their job. This relief means less gas is burned in those heaters, and that's where part of the savings comes into play. Since the water heaters depend on their thermostats to fire up the burners when the desired temperature falls below normal, we try to fool the heaters into believing they are already satisfied. This is done by supplying water heated by the cogeneration system's engine coolant to the intake side of the heater. If the heater's thermostat is set to fire the burners at say, 120°F, then the cogenerator supplies hot water to the heater above that temperature, say 121% F. This keeps the water heater satisfied and keeps the burners off, saving fuel.

Some facilities use once-through systems, while most facilities use continuous-flow systems where the hot water is circulated throughout the facility and back to the water heater. If no hot water is used, there is usually some temperature drop due to radiant- or conductive-heat loss, so the water coming back to the heater is colder than desired and must be reheated. Obviously, if hot water is used, then cold water make up is supplied from the city's supply and must be heated to the desired temperature. In any event the cogenerator is constantly using its engine-supplied "waste" heat to heat the facility's water needs.

There are times, however, when the heat load falls below the heat supply and the cogenerator must do something with this extra thermal energy. If allowed to build, it will simply overheat the engine; and the engine's protective system will shut it down. If this happens, the electricity supplied by the cogenerator is also curtailed. This may not be the most economical way to operate the system since electricity is needed within the facility virtually twenty-four hours a day while hot water may be a cyclical need. Hence, a method to keep the cogenerator running and supplying electricity while the thermal needs are low is to dissipate the excess heat through a radiator. This device is very similar to the radiator in an automobile and is referred to by many names: dump radiator; blow off radiator; heat dissipating device, etc. This radiator is located near the cogenerator, and the engine coolant is piped through the radiator. If the temperature of the coolant rises to a preset temperature, the radiator's fan will start and the heat will be blown off to atmosphere. As the coolant temperature subsides, the fan will shut down. This method of controlling the balance between thermal and electrical needs of the facility is well within the bounds of the regulatory agency's monitoring Qualified Facility (QF) operations. As discussed in Chapter 2, the annual system efficiency is monitored by FERC, and as long as it meets the 42.5% overall system efficiency, using blow off-devices is perfectly acceptable.

Other logistical considerations utilized in placing a cogeneration system into operation are security, safety, neighborliness, and common sense. The system should be secured against tampering by unauthorized persons. Even though the main engine-generator compartment is often enclosed in a weatherproof, locked cabinet; external pumps and controls could be tampered with if not secured. Safety is a consideration in that many of the pipes going in and out of the cogenerator enclosure are hot and can cause burns if the area is not secured.

Logistics of Installation   49

Neighborliness is simply realizing that with any engine some noise will emanate from the enclosure, through the exhaust if nothing else. While many small-scale cogeneration packages are sound attenuated, airborne noises can be a problem in a residential community during night-time hours when most normal activity subsides. Exhaust mufflers are commonly used, and the exhaust is pointed away from the nearest neighbor's house. There is no reason why a small-scale cogeneration system installed in a municipal swimming pool or a hospital cannot meet neighborliness criteria.

Common sense is often overlooked when systems are recommended to be placed on roofs without checking load-bearing needs; nearby potential fire sources like propane storage tanks or where exhaust fumes might waft into a kindergarten class. Otherwise, the logistics of installing a cogeneration system are driven by economics and good sense.

Figure 7-1 shows a typical installation of two 75-kW cogeneration

Figure 7-1. Small-scale cogeneration is provided by packaged modules such as this. Not only does the use benefit from low-cost electricity, heat can be recovered from the engine or turbine to further reduce energy costs.

plants. Note the clean lines of each package where all major components are enclosed in sound-proof cabinets for security and noise attenuation. The control panels are mounted outside the enclosures in their own cabinets to shield them from the internal heat of the engine inside the main enclosure.

Chapter 8

Permitting Requirements

Three agencies must be satisfied before a cogeneration system can be put into operation. They are:

1. The utility to which the cogenerator will be interconnected.
2. The city in which the cogenerator is located.
3. The Federal Energy Regulatory Commission, which will assign a Qualifying Facility (QF) number to the project.
4. The local Air Quality Management District

UTILITY REQUIREMENTS

Utilities vary in what is needed to inform them of the placement of an auxiliary-generating unit connected to their system. Some simply require notification and size of system. Others require elaborate paperwork in the form of an application; an engineering review; a metering review; a site inspection; and then a "client-utility-supplier kick off" meeting where the particulars of the above are discussed and a draft contract is given to the client explaining the utility's position in having this auxiliary source of power generation connected to its lines.

The size of the cogenerator does not seem to dictate the interconnection requirements of the utility. A 10-kilowatt system gets the same attention as a 1-Megawatt system.

Some of the information required on a typical utility application is:

1. Name and address of site owner and generator owner with contact names and phone numbers.

2. Engineering consultants or technical representatives of the project. Basically these names are the manufacturer and sales people of the cogenerator system.

3. Site information, such as whether the property is owned or leased and if the latter, how long is the lease. The site address is asked for.

4. A site map of the installation showing location of metering stations, rights of way and location of the cogenerator.

5. Is the customer residential or nonresidential?

6. Account numbers of the meters into which the cogeneration system will be tied.

7. What are the current facility operating hours?

8. How many hours/week will the cogenerator operate?

9. Will the cogeneration system operate in parallel with the utility or be isolated from the utility?

10. Will power be sold back to the utility?

11. What is the capacity of the generator in kilowatts; what is the voltage?

12. How many kWh per year will be produced?

13. How much power will be sold back to the utility?

14. Will the generators be FERC qualified?

15. Describe the project including electrical and thermal needs and outputs.

16. What is the prime mover?

17. What fuel will be used to power the prime mover?

18. Will an alternative fuel be used?

19. What natural-gas pressure will be required at the site?

Permitting Requirements

20. What quantity of natural gas will be needed to power the prime mover?

21. Give an estimate of the waste-heat recovery of the cogeneration system.

22. What will the waste heat be used for?

23. Give the size of the generator, the number of generators, the manufacturer of the equipment, the voltage, and the model of the unit.

24. When will FERC approval be given?

25. When will construction begin, when will it end, when will testing be required, and when will project be fully energized?

26. Has financing been procured?

27. Have air-quality permits been applied for?

28. When will all permits be procured?

29. Advise whether a preliminary or detailed interconnection study is required.

After this application has been submitted to the utility, the utility will prepare an interconnection contract for the client to sign. This contract will outline the utility's responsibility to provide back-up power to the client in the event the cogenerator is down for maintenance, either forced or planned. It will also define the length of the contract. Here utilities are prone to shorten the actual contract time but advise that the contract may be renewed for 30 or more years. This is a method by which the utility is offering itself an out to provide back-up power and possibly make the cogeneration system's production output not a part of its "running reserve" power. Since any attempt by the utility to curtail back-up power flies in the face of the PURPA law, these clauses are mostly scare tactics by the utility and would be difficult to enforce. Clients have been known to object to such clauses and the utilities have tempered their stances.

The procurement of gas will be addressed by the utility if it also supplies the natural gas that will be used to power the cogenerator. If the natural gas is to be procured from a third party, not associated with the electric utility, this contract will not be in the electric utility's purview. Most, but not all, natural-gas providers will offer a price for gas that is equal to what the electric utility pays for natural gas to power its boilers for steam to electric production. If the utility is also the gas provider, a separate gas-procurement contract is tendered to the client. Gas provided to a cogenerator under this type of contract is referred to as non-core gas and is priced under that price schedule. The utility will show the various prices it pays for a) procurement, b) transportation, and c) distribution of that gas to the client. If a client should choose to procure its own supply of gas, which is certainly in the client's right to do so since the deregulation of natural gas, the utility would still distribute that gas and charge the client accordingly.

The difference in price for noncore gas versus core gas is quite substantial. Generally, that price difference will be 50% of the core-gas price. So, if a core customer is paying $0.65/therm, the noncore customer may be paying $0.32/therm. This is a substantial savings in the overall Cost/Savings analysis that was discussed in Chapter 6, and it behooves the client to make sure he is getting the best "utility" price for natural gas that he can.

Most of the rest of the standard utility contract for interconnection is boiler plate that requires Public Utility Commission approval to alter, is neither advantageous nor deleterious to the client, and may be dispensed with summarily.

An engineering charge is often made by the utility to cover its cost of investigating the cogeneration interconnect. There is also some hardware costs that the utility will impose such as the cogeneration output meter that shows the utility just how many kilowatts per month the system generates. A gas and electric utility will use the kilowatt output to calculate the natural gas that should be supplied at noncore rates based on the same efficiency as the utility's power generation. This is expressed in Btu's/kilowatt and will be in the 9,500-to-10,700-Btu/kilowatt range (often expressed as 0.095 to 0.107 therms/kilowatt). So, if a cogeneration system produces 60,000 kilowatts in a given month the utility will multiply that production by its efficiency factor to determine the gas that will be sold at noncore, or cogeneration, rates. Sixty thousand kilowatts times 0.095 therms/kilowatt = 5,700 therms that will be

Permitting Requirements 55

priced at noncore gas rates that month. The cogenerator will use more gas than that because its efficiency in converting the natural gas it uses to electric energy will be less than the utility's. Unfortunately, the utility does not take into account the transmission and distribution losses when applying this factor.

Nonetheless, the excess gas that is consumed by the cogenerator is priced at the next tier of Non Core gas rates which are only slightly higher than the best Non Core rate. If the client's overall gas usage is 90% hot water heating, then only 10% of the client's gas bill will be priced at the full Core rate while the remainder of the gas usage will be priced at various Non Core rates. The overall affect is a considerable savings in gas cost to the client that is operating an on-site cogeneration system.

It is important for the cogeneration user to be familiar with the Core and Non Core rates and use that information as a negotiation tool with whomever is supplying the natural gas. Appendix V shows typical utility gas rate schedules for both commercial and cogeneration gas supplies, Core and NonCore.

If the prime fuel used by the cogenerator is diesel oil and the utility is operating on oil, the oil supplier is not bound by the rules that have been established when natural gas is the prime fuel and no standards exist. Obviously, if the utility is using coal as its prime fuel source, and the cogenerator is using natural gas, it further behooves the client to negotiate his best deal with the natural gas supplier using this information.

CITY REQUIREMENTS

Here, the requirements are similar to any facility improvement. The City wants to know what is being installed: does the installation meet existing city codes with respect to wiring and plumbing, and are the components safety inspected and bear Underwriting Laboratory (UL) or Engineering Test Laboratory (ETL) stickers?

Application is made to the city inspection department, and their inspectors will visit the site when construction is complete to issue its permits. Fees are charged for this service and are generally included in the overall cost quoted to the client by the cogenerator manufacturer or sales representative. These fees range from $50 to $500 depending on the city and the scope of the work.

The Utility will not act on its interconnection contract without the city's inspection being completed and permits issued.

Recently, cities are requiring more information on the type of equipment being installed, it's adherence to UL or equal conformation, noise levels, temperature emissions, safety considerations such as insuring exhaust stacks are above adjacent roof lines, and overall aesthetics of the installation if it's outdoors and subject to viewing by neighbors or passersby. On an actual project a vendor was installing an attractive microturbine cogeneration package alongside a customer's building that faced the street. The local city permitting department insisted that a fence be built in front of the unit and that shrubbery be placed in front of the fence. It was pointed out that an ugly utility transformer in need of a paint job was also facing the street but the city replied that that was OK.

Design review is a new buzz word being applied to new construction projects. This can cost thousands of dollars and slow down the issuance of a permit by months if not addressed professionally and early in the permitting process. It behooves a vendor or developer to ask the city where the new equipment is to be installed what their criteria are even before an order is issued by the client. The reason for this is that the client has already been through his round of permitting for other projects and knows that this can be a fatal stumbling block in allowing the project to proceed.

FERC-QUALIFIED FACILITY PERMIT

Application is made to the Federal Energy Regulatory Commission for a QF number. This is a straightforward application stating what is being installed and the energy being produced, both electrical and thermal. The following is an actual letter to FERC asking for the QF number. FERC has replied and affixed QF number 97-97-000 to this project: (Figure 8-1)

A copy of this letter is returned to the client with a QF number stamped on the letter. This letter is given to the utility to confirm compliance with FERC, which sets the PURPA law into motion as it pertains to utility interconnect, gas-contract pricing, and overall cogeneration interface with existing national interests. It is rare if this permit requirement should cause any difficulty in implementing the cogeneration project.

Figure 8-1

KOLANOWSKI & ASSOCIATES

COGENERATION ENERGY SPECIALISTS

7221 Linden Terrace
Carlsbad, Ca. 92009

Phone 760-431-0930　　　　　　　　　　　　　　　Fax 760-431-0955

May 2, 1997

Office of the Secretary
Federal Energy Regulatory commission
825 N. Capitol Street, N. E.
Washington, D. C. 20426

Re:　Notice of Self Certification for Qualified Cogeneration Facility

1. Owner and Operator of the Qualified Facility:
2. Description of the Qualified Facility:
    a. Facility uses a reciprocating engine with appropriate heat recovery equipment (Topping Cycle).
    b. Electric Output Capacity in Kilowatts:　　　　　　　　　　120
    c. Useful Thermal Energy Output in Btu/HR:　　　　　　6,381,000
    d. Fuel Input in SCFH (at 1020 Btu/SCF HHV):　　　　　1064.41
    e. Facility will be located on site at:
        Jewish Community Center
        4126 Executive Drive
        La Jolla, CA 92037
    f. Projected maximum on line annual run hours:　　　　7800
    g. Annual Efficiency Calculations (MBtu):
        Total Fuel in HHV:　　　　　　　　　　　　　　　　8,302
        Useful Power Output:　　　　　　　　　　　　　　　3196.4
        Estimated Useful Thermal Energy Output:　　　　　4976.4
        PURPA Efficiency:　　　　　　　　　　　　　　　　68.47%

3. Primary Energy Source:　　　　　Natural Gas
4. Primary Energy Supplied by:　　　San Diego Gas & Electric
5. Percentage of Utility Ownership:　0%
6. Expected Date of operation:　　　June, 1997

Enclosed is the original application and 1 copy with a Self Addressed Stamped Envelope. Please return copies stamped with the QF Docket Number and the Date of Filing to me.

Sincerely,

Bernard F. Kolanowski

The question of air quality was touched upon earlier, and this may be another permit requirement in certain air-quality districts. The California South Coast Air Quality District is the most famous district in the country when it comes to protecting the environment. They have been in business the longest and have established rules for allowing or disallowing projects that need to comply with those rules. In the application of cogeneration systems powered by natural gas this district has been very reasonable in the knowledge that this technology effectively increases fuel efficiency and that overall less pollution permeates the atmosphere. However, they still require Best Available Control Technology (BACT) and will impose the need for catalytic converters on the engines that power the cogeneration system. This is generally not a problem although it does add an additional cost to the cogeneration system.

The developer or client should be aware of the local Air Quality District requirements before proceeding with the project. In the vast majority of the cases where small-scale cogeneration is involved, the districts exempt engines with horsepowers less than 200 to 500 hp depending on the location. In that case, any 120-kilowatt system will receive an exemption.

AIR QUALITY MANAGEMENT DISTRICT PERMIT

Every state and locality has concern with the air quality issues when new fossil fuel burning equipment is to be installed. The main concern seems to be NO_x production but the quantity of CO and unburned hydrocarbons are also addressed. If sulfur laden fuels are being burned, then the amount of SO_x emitted is also addressed. Catalytic Converters are often used to limit NO_x production and may have to be applied depending on the limits of the local district. Often, size of prime mover is the determining factor used in when controls must be imposed. Two hundred horsepower and under are exempt in areas where pollution has not reached the non-attainment criticality, while 50 horsepower may be the upper limit in most non-attainment areas.

Cities will not issue a permit to construct unless they have evidence that the Air Quality District requirements have been met and an appropriate permit is produced.

Chapter 9

Operation & Maintenance

Operation of small-scale cogenerators is automatic. Whether the cogeneration system is intended to run 24 hours/day or within hourly parameters, it will be sensitive to preset limits on time, temperature, or internal operating parameters. Of course, the system may be started and stopped manually or via computer commands entered manually, but usually that's only done when the system is restarting after a manual or forced shut down.

Let us look at these variations of control and operation:

Time

A facility, say a commercial laundry, opens its doors at 6:00 a.m. and uses the cogeneration system to provide hot water and electricity. It may set the cogeneration system to start up an hour before normal facility operations in order to insure the hot water production is up to specified temperatures in the storage tanks. Any electricity produced during this "pre-operative" phase may be given or sold back to the utility presuming the electrical load in the facility is very light.

When full facility operation is underway, the cogeneration system is producing hot water and electricity to meet the needs of the facility per its design parameters. Design parameters mean that at any given time, the constant production of the cogeneration system may fall short of the facility's needs, in which case standby systems, i.e. the facility's hot water heaters and or the public utility's electricity, will be fed into the system automatically. Should the cogeneration system's output exceed the facility's needs in hot-water temperature,, an automatic bypass will engage causing excess heat of the hot water to be exhausted via a dump or blow-off radiator while the electric production will remain constant. Should the cogenerator's electrical output exceed that of the facility's needs, that electricity will be given or sold back to the utility. All this happens automatically with the cogeneration system and facility systems set up to accomplish these variables.

When normal shutdown of the facility is planned for 10:00 p.m., the timer system of the cogenerator will kick in and shut the cogenerator down at a prescribed time, possibly as soon as a half hour or so after official closing time. Should the facility decide to extend or shorten its hours of operation, the manual control of the cogeneration system can override the timer controls and either keep the cogenerator running or shut it down sooner than planned.

Temperature

Smaller cogeneration systems of up to 30 kW may prefer to operate without a blow-off or dump radiator and use thermostatic controls to shut down the cogeneration system if the hot-water temperature should exceed preset maximums due to lower hot-water usage. In these kind of systems the hot water is often produced at temperatures in excess of the facility's needs and is stored in hot-water storage tanks. Mixing valves will regulate actual facility needs by introducing cold water to the elevated hot-water temperatures to produce the required "mixed" temperature for the facility.

This method allows the cogeneration system to continue to operate during periods of low hot-water usage, producing electricity, while the temperature of the water rises inside the storage tank. If the temperature rises to a preset maximum where danger to the cogeneration system may occur, the cogenerator will automatically shut down. The facility will then be drawing its electrical needs from the utility grid. When the temperature in the storage tank subsides to a preset value, the cogenerator will automatically start and continue to run until again the needs are exceeded.

This kind of control is practical when knowledge of the facility's needs are well known so that too many shut-downs are avoided during normal operating times. Of course, a timer system may also be used if the facility also has less than a 24-hour/day operational time.

By not incorporating a blow-off radiator, the user of the cogeneration system avoids having to report FERC efficiency results to the utility annually. It is taken for granted by both FERC and the utility that all the produced heat is used by the facility, and the electricity is fully used whether by the facility or the grid.

Internal Operating Parameters

The cogeneration system will be instrumented to respond to its own operational and safety needs. Typical of the instruments that are employed in cogeneration systems are:

- Low oil pressure
- High coolant temperature
- Low coolant temperature
- High output voltage
- Low output voltage
- High or low frequency
- High generator temperature
- Low gas pressure (fuel)

The utility to which the cogeneration system is connected will be particularly interested in the system's ability to shut down on high- or low-voltage output and high-or low-frequency output for safety reasons to their own grid conditions. They often require a demonstration of the system's ability to shut down when those conditions occur before official interconnection permission is given. This is referred to as the "trip test." It is conducted with an electric-voltage inducer called a variac that is connected to the output wires of the cogenerator while the system is down. Low and high voltages are simulated to see whether the contactors trip shutting the unit down.

Monitoring of the cogeneration system's operational parameters is often done automatically and remotely via a dedicated telephone line. If a shutdown occurs, the system is set to call the monitoring station alerting it to the shutdown, usually giving the reason the system shut down. The monitoring station has the option of evaluating the reason for the shutdown, restarting the system manually to see if the shut down was due to an anomaly and then watching closely to see if performance is normal, or shutting the system down in a lock-out condition and dispatching a service supervisor to the job site. Monitoring may be done by the company under contract for performing the maintenance of the system or by the user of the cogeneration system or by both.

MAINTENANCE

Maintenance starts with good design and good quality assurance in the factory and good installation parameters in the field. That is true of almost any mechanical and/or electrical system.

An example of how design can affect maintenance is found in the way the engine is attached to the generator. In small systems, 10 to 30 kW, and especially in the early days of small-scale cogeneration, this attachment was made by a distinct coupling between the engine and the generator. They were commercially procured and incorporated self-aligning and shock-absorbent features to maintain integrity of alignment and absorption of thrust. Both these features would affect the bearing systems of the engine and the generator if allowed to become misaligned in the vertical or angular direction or if excess thrust was produced in the horizontal direction. Commercial couplings accomplished these safeguards but exposed themselves to wear and eventual replacement. It is not unusual to replace couplings every 4,000 hours of operation.

A better design was to attach the generator to the engine by extending the engine shaft to be integral with the generator and using a ring flange on both the engine and generator to mate those two elements together. This totally eliminated the coupling and assured alignment in the vertical and angular directions. Thrust bearings in each component took up any horizontal thrusts. This feature eliminated the maintenance associated with couplings as well as the expense of replacing the couplings periodically.

Good quality assurance, or quality control, at the factory level insures that good designs are produced with accuracy. There's no sense in having a good design like the mating-ring flanges if those flanges are machined incorrectly thus causing stress on the engine shaft, which destroys the bearings.

But, notwithstanding good designs and good quality control, every mechanism with moving parts will require maintenance, both preventive and forced. Preventive maintenance is the periodic changing of oil, filters, spark plugs, seals, and coolant. Forced maintenance is the replacement of engine pistons due to overheating or abrasion causing an unexpected failure and subsequent shutdown of the system. Both types of maintenance will require shutdown of the system, but preventive maintenance will reduce the periods of forced maintenance and can be done during optimum hours. In cogeneration work, optimum hours are when the utility is not charging peak prices for electricity. Peak electrical prices are generally in the late spring through summer and early fall months when heavy electrical loads are being experienced, especially during afternoon hours. During those months, preventive maintenance must be done in early morning or evening hours to preclude a planned shutdown of the system. Obviously, preventive maintenance is important during these periods to avoid forced shutdowns at inopportune times.

Any heavy maintenance such as engine overhaul should be done before electric utility peak-price periods begin.

MAINTENANCE CONTRACTS

Most users of cogeneration systems will prefer to have the provider of the system do the maintenance under a maintenance contract. The two most common types of maintenance contracts are "lump sum in advance" and "production sums in arrears." Both types of contracts encompass not only preventive maintenance but also include replacement of any component or part in the entire cogeneration system that was provided under the sales contract. This is referred to as an "extended maintenance and warranty" contract. In each case, monitoring of the system remotely is often included in the contract..

Lump Sum In Advance

This type of maintenance contract stipulates a lump-sum payment that varies with the cogeneration system size. The larger the system, the higher the maintenance cost. One manufacturer quotes prices as such:

kW Size Cogeneration System	Annual Maintenance Cost
40 kW	$6,630
60 kW	$8,970
120 kW	$11,000
250 kW	$17,500

Payments for this type of contract are usually made annually or quarterly in lump sums.

Production Sums in Arrears

In this type of maintenance contract, the payment will be made by the client based on the production of the cogeneration system. Normally it is based on the kilowatts produced in any given period. Sometimes it will be based on the hours the cogeneration system ran during a given period. In most cases the period in question will be a month. Therefore, if a 120-kW cogeneration system runs for 650 hours during the month of July and produces 78,000 kW, the client will pay a maintenance fee of $0.016/kW, or $1,248 for the month of July. That payment will be due on or about August 10. The same format holds true if the payment is based on hours of operation and the $/hour will vary with the size of the machine, while in a kilowatt-production-based maintenance contract, the $/kW may remain the same because the power production will vary with the size of the machine.

Often, the "production sums in arrears" contract will result in slightly higher maintenance fees annually than the "lump-sum" contract if the cogeneration system runs virtually perfectly during the year. For instance, in the above example, the customer would be paying $1,248 per month (12 months = $14,976 per year or $3,976 more than if they had paid "lump sum in advance." However, the customer gets the satisfaction that if the machine only ran 60% of its allotted time and produced only 60% of its maximum power, they would only pay $8,985 for the year rather than the $11,000, whether the machine ran or not. Also, the time-cost factor of money comes into play with the "lump sum in advance" contract versus the "production sums in arrears" contract.

The customer may or may not have a choice as to what type of contract is offered him, but if he knows the differences, he may be able to influence the type of contract he wants and, therefore, gets.

One last issue is that at times a client may forsake the extended-warranty facet of the maintenance contract and opt to pay a time-and-material cost for both preventive maintenance as well as for replacement components and parts. In this case the client and provider may agree on a flat monthly fee to pay for the preventive maintenance and monitoring facets of the system, but then have the client pay an hourly cost for any work done in parts and component repair or replacement as well as the cost of the parts in question. These fees are negotiated between client and provider.

Chapter 10

Pitfalls of Cogeneration

There are a number of pitfalls that will cause a cogeneration project to be unsuccessful. If success is measured in a project's ability to meet the proforma expectations in overall utility savings and return on investment, then anything less than that will render a project unsuccessful. However, partial success still may be acceptable. For instance, if a project pencils out to have a return on investment of 35% and attains only a 28% savings overall, that project still returns greater than normal returns compared to other investments and can be deemed successful, even if it does not meet initial expectations. Hence, the primary concern is when a project fails abjectly. When the returns are considerably less than what can be earned in conservative investments, then the project would be said to have failed.

LACK OF RUN TIME

The first pitfall is lack of run time which is attributed to failures caused by design shortcomings, maintenance shortcomings, or the combination of both. If the design of the system causes innumerable shutdowns and the vendor makes every attempt to correct those design shortcomings, the project will probably turn itself around and become successful. It is when those efforts reach a stone wall and neither the vendor nor his major suppliers can solve the problem or simply do not have the resources to continue with the solutions and so simply walk away from the project.

That very problem, lack of resources, has caused much of the consternation surrounding the efficacy of small-scale cogeneration. There have been too many vendors that, after a few years of operation, have been unable to capitalize their company sufficiently and have ceased to exist leaving the users to fend for themselves in finding qualified maintenance outlets and solving major design shortcomings. Chapter 13 discusses current and past vendors and will shed more light on this major pitfall of small-scale cogeneration.

Therefore, if a system fails to operate for the required number of hours it was designed, the overall savings will be less. The most common reason for run-hour shortfall is not design shortcomings but maintenance considerations.

One such company with which the author has had direct experience, manufactured units in the 10- through 30-kilowatt sizes. These units were well designed and manufactured and were given test runs before being installed at the customers site; and virtually all installations included an "Extended Warranty and Maintenance" contract. That meant that the manufacturer-vendor entered into contract to supply all routine maintenance and replacement parts and components at no other charge than the client was paying in a production-arrears contract. In this case it was based on number of run hours per month. A service technician was scheduled to visit all installations at least once a month routinely. No installations had any telemetering-of-performance functions, operational proof or automatic feedback. It was up to the client to inform the company of any forced downtime. However, because the units were often installed on roof tops and other "out of sight out of mind" places, the client did not know when a unit might be down. The units had hour meters incorporated into their design, which was the basis for the maintenance contract billing.

Often, a technician would visit a job site to find the cogenerator not operating and learn that the number of hours of run time for that month was just a percentage of the previous month's total. Yet, the unit did not appear to have any obvious repair problems. What had happened was that the unit ran out of oil and was automatically shut down by the oil pressure sensor to prevent destruction of the engine due to lack of lubrication.

It was determined that as a unit aged, its oil consumption increased considerably and the oil reservoir, usually only that amount contained in the crankcase, was insufficient to last a full operating month before it was depleted and the unit shut down. Not only did the client not receive the anticipated utility savings and economic benefits, but the manufacturer-vendor did not receive its full maintenance payment because of lack of run hours. One solution was to dispatch service technicians more frequently to those job sites until a more permanent solution, i.e. increasing the oil-reservoir capacity, was found. Unfortunately, that manufacturer went out of business before that design modification was made, but subsequent manufacturers learned from those mistakes and included expanded oil reservoirs plus telemetering in their designs.

RATE CHANGES

A second pitfall that may occur in a cogeneration project's ability to meet its proforma design in economic savings is when the utility changes its rate structure and or its rates. Rate structure is when an electric utility changes from a straight-energy-charge rate based on cents per kilowatt to a structure that includes both energy charge and demand charge. When this occurs, the energy charge usually is reduced substantially, and the demand charge is imposed based on a kilowatt usage in any given 15-minute period. While the overall cost to the client for electricity may not be significantly different, the savings that the cogenerator will produce may be affected if the cogeneration system goes down during the peak-demand time for 15 minutes or more. If that should occur, the client is billed for the total maximum monthly demand and gets no credit for the kilowatt rating or production of the cogenerator for that entire month, even if the cogenerator were only down for a half hour out of the entire month.

Utilities are writing their interconnect contracts in such a way that they may be worded that they're year-to-year contracts. Therefore, if a client starts out with an energy-only rate structure, the contract may be changed after a its anniversary date to a time-of-use contract that now includes demand charges. These kinds of contracts put an even greater demand on not only preventive maintenance, to insure the system will run through peak demand times, but also on the timing of maintenance so that it does not occur during peak-demand times requiring the unit to be shut down for inspection, oil changes, and other preventive-maintenance chores.

Outright rate changes may also affect the economic savings of a cogeneration system. Should a utility reduce its rates in its service area, the cogeneration user will still get the benefits of having displaced a certain number of kilowatts he would have purchased from the utility, but now those kilowatts would be at a lower rate than when he first purchased his cogeneration system. Rate reductions are not very common, but with the deregulation of electricity becoming more popular and customers having the ability to choose who their electricity supplier will be, savings will occur in those areas. As of this writing it is uncertain just how much the deregulation of electricity will affect overall kilowatt costs, but at present, 2 - 3 % seems to be the number most independent producers are quoting.

FUEL-RATE STRUCTURE

Fuel-rate structure will normally not change during the life of a cogeneration system. However, this section may be a proper time to talk about how fuel rates may affect a cogeneration project from the beginning. While PURPA required utilities to allow cogeneration systems to be interconnected with them for purposes of standby power and sell-back provisions, it did not necessarily grant cogeneration systems the same fuel rate as the utilities themselves. However, in many parts of the country gas providers have sold natural gas to cogeneration installations at the same rate as it sells to the utility. Other gas companies will give the cogeneration client a large-customer rate, which may approach that of the utility. Yet other gas companies may simply negotiate a rate with the cogeneration client. Some gas companies will sell their gas at the same rate as the client qualified for in the first place, regardless of the cogeneration system in place.

The cogeneration system will cause an increase in the amount of gas used by the facility since the fuel is used for both heating and electrical generation, but there will be an offset in the gas that had previously been used for heating water via the client's in-house water heaters. The overall gas sales in the area will be less due to the higher efficiency produced by the cogeneration system. Nevertheless, there is no universal cogeneration gas rate available throughout the country. In comparing the "cogeneration" gas rate from a California gas supplier to that of a Virginia gas supplier, the difference was about $.09/therm, with the Virginia supplier calling the rate "Large General Service," and was higher than the California rate.

Suffice it to say that the fuel costs go into the equation when a cogeneration system is being analyzed for economic savings, and unless the gas supplier decides to change the classification of the client during the life of the cogeneration system, the initial rate structure will not change.

Other fuels that may be used in the cogenerator's engine drive may be diesel oil, propane, and methane. Few of these fuel suppliers have special rates for cogeneration systems, and negotiated rates will prevail based upon annual usage.

POWER FACTOR

The displacement of kilowatts by use of an on-site generator may be significantly affected if the power factor of the facility is low. Power factor is the term applied to the efficiency with which supplied power is used within the facility. A high power factor means that electrical power is being utilized effectively, while a low power factor indicates poor utilization of electrical power.

Therefore, power factor is a measure of the real power-producing current in a circuit relative to the total current in that circuit. It indicates how much real work is being done relative to the total amount of current drawn by an electrical device.

Low power factor in a facility means that the supplying generator, whether it is the central utility's generator or the on-site (co)generator must put out more power in the form of kVA, or kilovolt amps, to accomplish the real work needed to be done within the facility. kVA power is referred to as Apparent Power. The real work is referred to as Real Power and is measured in kilowatts (kW). The relationship between Apparent Power (kVA) and Real Power (kW) is displayed in a Power Triangle as shown in Figure 10-1. The angle between kVA and kW is called theta (Θ) and is the power factor percentage of the facility. (See Figure 10-1)

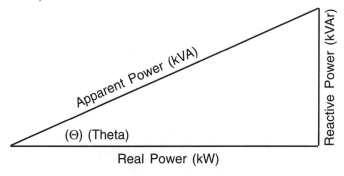

Figure 10-1

The other leg of the triangle opposite theta represents Reactive Power (kVAr). This power serves no useful function, but is an indication of the drain on your power supply that is performing no useful benefit to the work at hand. If the power factor were unity (1.0), there would be no Reactive Power (kVAr) and therefore, no drain on the power being supplied. Where the electrical power being supplied is being used for lighting or heaters, which are resistance loads, the power factor would be unity. However, washing machines, compressors, refrigerators, pumps, all of which use induction motors to drive them, have inductive loads and usually run at lagging power factors of 0.8 or less. These devices will drag down the overall facility power factor, including that power used to supply lights or heaters.

What this means in cogeneration work is that when a cogenerator is installed in a facility with a low power factor, the generator must produce more work to maintain the same kilowatt displacement advertised in the contract. Usually, this is not always possible, and what happens is that the generator in the cogeneration machine will produce the advertised current in amperes, but only a percentage of that current will be available for useful work. The result is a decrease in the displaced kilowatts. A client may then argue with the cogenerator supplier that the cogenerator is not putting out the advertised kilowatts. The cogenerator supplier may counter by increasing the output of the cogenerator, which increases the current, but only at a sacrifice in fuel consumed, which would be higher than the contract initially showed. The customer may then complain about the fuel usage's (which he pays for) not being in line with the contract.

In either case the savings to the client will be less than advertised.

If the utility imposed a power-factor penalty to the client reflective of the kVA it needed to produce to supply the client with the needed kW, then the client's cogenerated power would offset that power that was purchased from the utility, and the power factor penalty would be less. That savings would then justify the cogenerator's lesser output as the combination of cogenerated power, and savings in power-factor penalty would increase the overall savings the cogeneration system is providing making for a happy client. Unfortunately, few utilities impose a power-factor penalty on loads that reflect the small-scale cogenerator's optimum clients. In other words, when the facility's overall kW usage is relatively small compared to the utility's total output, power-factor penalties are usually ignored.

However, low power factor hurts the client in other ways, too. The current flowing through electrical system components, such as motors, transformers, and wires, is higher than necessary to do the required work. This results in excess heating, which can damage or shorten the life of those components and, in extreme cases, even cause fires. Low voltage conditions may also prevail, which result in dimming of lights and sluggish motor operation, especially the starting of electrical motors where starting-current requirements are 5 to 6 times that of normal operating current.

Correction to a facility's low power factor can be accomplished by the use of capacitors added to the facility's power-distribution system. They act as reactive current generators, which helps offset the inductive current devices in the system, thereby increasing the overall power factor in the facility.

The pitfalls of cogeneration are not something that should thwart the investigation and eventual inclusion of this type of energy conservation into a responsible client's thinking. The benefits in economic savings, conservation of fuel savings, and ecological savings should be included in the new words being coined today - Green Power - as beneficial as wind power, hydro power or geothermal power. Careful attention to vendor selection, machine design, utility attitude and facility capabilities will overcome most, if not all, of these pitfalls and make cogeneration a viable method of energy conservation.

Chapter 11

Financing Cogeneration Projects

When a client decides to employ a cogeneration system on its premises, he has a number of choices as to how to effect that employment:

1) **Purchase.** The client can purchase the system outright at the vendor's quoted price and terms. The client becomes the owner of the equipment and can operate and maintain it as he sees fit. As discussed in an earlier chapter, the maintenance of the equipment can be contracted out to a separate company at a stipulated cost and may even include an extended-warranty provision where all replacement parts and components are included in that stipulated cost.

2) **Leasing.** The client can lease the system for a period of three, five, or seven years paying a leasing company the requested amount on a monthly basis. When entering into this kind of financing arrangement, the client has choices on how he may want the lease structured. First, he may want to show all lease payments as an expense and write them off against taxable income. This is often referred to as a "true lease" and must include some buy-out provision at the expiration of the lease period. The Internal Revenue Service has guidelines as to how a true lease should be structured to allow the write-off provisions. Usually they require that at the end of the lease period the purchase price must be "fair market value." Normally, 10 to 15% of the original-equipment price would be considered fair market value. Some leasing companies will ask for this percentage up front, so the final buy out will be for no additional money.

Secondly, a lease may include provisions for extending the lease at the end of the initial period. Re-leasing is a very profitable experience for the leasing company, but it does provide an option for the client. This is fairly common if the client opts for incorporating the latest design modifications to the equipment thereby updating the system and then opting to continue, or renew the original lease arrangements at re-negotiated terms.

Lastly, the lease may include provisions to have the equipment removed from the premises by the leasing company, forfeiting the need for the 10 to 15% buy out provision. The client would then have no additional obligation to the leasing company nor the original vendor.

Leases may be tailored to reflect smaller monthly payments with a balloon payment due at or near the end of the lease. A large down payment may be made initially, with smaller monthly payments and no buy out payment requirement. Often, with equipment that provides a known potential savings to the client's operating budget, a lease may be tailored so that the monthly savings equal the lease payment causing the client to suffer no additional budget expense, while enjoying potential tax benefits. Of course, the risks of attaining those monthly savings lie entirely with the client, and any shortfalls are made up by the client to the leasing company.

Creditworthiness of the business is the hallmark of tailoring a lease to suit a client's needs. A leasing company will work any reasonable plan with a company or business that has shown an ability to perform profitably over a long period of time or has an acceptable track record of meeting its obligations in the case of a nonprofit organization.

Leasing provides various advantages to both the lessor and the lessee. For the lessor, advantages include availability of accelerated depreciation to reduce tax liability, applicable investment and energy tax credits, and the residual value of the equipment. For the lessee, advantages include 100 percent financing (no capital requirements); the possibility of lower payments than a bank loan; off-balance-sheet financing; and no direct reduction in the lessee's borrowing capacity.

3) **Government funding.** State-assisted financing may be available for energy conserving projects. California, for instance, has set up a program under the acronym SAFE-BIDCO. This stands for State Assistance Fund for Enterprise Business and Industrial Development Corporation.

The eligible borrowers are small businesses and nonprofit organizations such as the YMCA, Jewish Community Centers, and municipalities.

Eligible projects include cogeneration along with lighting changes, weatherization, energy management systems, and just about any qualified project that will meet their energy-conservation standards.

Qualified borrowers will receive up to $150,000 in loans at 5% for five years. The project must have a return on investment of ten years, or less, and be audited by an independent auditor to insure the project meets the standards SAFE-BIDCO has set for the specific type of project. Loans may be combined with other financing when the project cost exceeds $150,000. Creditworthiness also goes into the equation for eventual project loan approval.

The unique aspect of California's SAFE-BIDCO program is that the funds for borrowing come from Petroleum Violation Escrow Account money, basically money provided by oil companies that overcharged the state for their products. The California legislature enacted this plan in 1986, and by 1996, SAFE-BIDCO had loaned out over $7,000,000 to California firms for energy conservation projects.

4) **Grants.** Various government agencies and support groups interested in attaining conservation of energy awareness will often grant money to implement cogeneration projects. Such grants require following a format to demonstrate energy savings before the grant is issued, but once issued requires no pay back of the granted money. The Gas Research Institute is one such group that has made grants to various individuals and companies to foster conservation of natural gas in demonstration projects that may lead to commercialization of the final product.

The Federal Energy Regulatory Commission, the Department of Energy, and state energy departments are a source of learning where such grant and or low-cost loan money is available for energy conservation projects.

5) **Shared Savings.** This is a plan whereby a third party financier will put up the entire cost of the cogeneration project in return for a share of the utility savings that accrue because of the cogeneration system. The client, or "host," has no obligation, no outlay of capital, and no risk whatsoever other than granting a space for the accommodation of the cogeneration system.

Shared-savings Agreements come in many shapes and forms. The basic concept is that when less electricity is purchased from the local utility and fuel to run the cogenerator is contracted for at less than prevailing rates, also contributing to less money spent, there is an overall savings. A financier will invest the money to install the system on the client's premises in return for a share of the overall savings. That financier must not only install the capital equipment, but operate the system as if it were his own, which, in fact, it is. The financier owns the equipment and therefore is responsible for all operating costs including the fuel to run the engine, the maintenance of the entire system, and possibly the insurance costs to cover his property while it is on the client's premises. The client, in turn, receives his share of the savings from a pre-arranged contract.

Some third-party financiers tailor the contract so that savings are shared after all expenses of operation are paid for, netting out the savings and passing on a share to the client. With those types of contracts, the share may be quoted as 50% or more. If the system suffers from some of the pitfalls such as requiring heavy maintenance resulting in less running time and added costs, there may be very little "net" savings available to share. The third-party financier does have a vested interest in seeing that the run time of his system is as high as possible in order to garner the savings, but his risk is ameliorated somewhat with a "net" savings contract.

The more fair Shared-savings Agreement is one in which the client shares in the gross savings of the system. This percentage may be quoted as little as 10% or as much as 30%, but it is a direct reflection of the kilowatts produced by the cogeneration system as well as the fuel savings. All costs of operation come out of the financier's share after the client has been paid his "gross" share.

In this type of Shared-savings Agreement, the contract is written in such a way that the client, or "host," purchases the electric power output and the thermal energy produced from the Third-party Provider at an established discount under prevailing utility rates. If the contract calls for a 20% discount from prevailing rates and the cogeneration system produced 200,000 kilowatts of electricity and 3,000 therms of heat and the prevailing rates are $0.07/kilowatt (energy and demand charges included) and $0.72/therm, then the client would receive 200,000 kilowatts times $0.07 cents = $14,000 × 20% = $2,800 in electrical savings and 3,000 therms × $0.72/therm = $2,160 × 20% = $432 in thermal savings for a grand total of $3,232 in revenue for that month alone. The remainder of the savings, i.e. $12,928 is paid to the Third-party Provider by the client (host) as if the Third-party Provider were a utility.

Without cogeneration, the client's utility bill would have been $16,160 for that month. With cogeneration, the client is only obligated to pay $12,928, hence a savings of $3,232 for the energy savings the cogenerator produced for that month.

The Third-party Provider uses the money paid to him by the client, $12,928, to pay for the fuel, maintenance, insurance and the debt service incurred when he purchased the system to install on the client's premises. A Third-party Provider will want to see a proforma that will return to him at least a 30% return on investment before entering into a Shared Savings Agreement with a client.

A typical Shared-savings Agreement reflecting this type of arrangement is shown in Appendix I.

Whichever method is used to finance a cogeneration system it behooves the client to look at his own criteria of putting his money to work to realize a return. In the case of cogeneration, the client can provide all the electricity and hot water he needs from the local utility and his own hot-water heaters. Therefore, there is no urgency to install a cogeneration system unless the return on his capital expenditure is sufficient to meet expectations that go beyond normal investment returns. Since normal investment returns of 8-12% can be attained with a moderate degree of safety, the cogeneration project that shows 20 to 30% returns must be favorably evaluated and implementation considered. Anything less than 20% should probably be rejected as not having enough safety margin in which to make the investment.

Conversely, a Shared-savings Agreement, that encompasses no risk on the client's part must be considered most favorably if the Provider has satisfactory credentials in this field and can point to references that presently enjoy his services. This is especially so if the Shared-savings Agreement offered is of "gross" savings where savings come off the top prior to expenses and debt service.

Chapter 12

Case Histories

The success of a cogeneration project is measured in terms of what it saved the client in utility costs after on-site operating and maintenance expenses are deducted from gross savings. Whether the facility is a nonprofit organization such as a YMCA, Jewish Community Center or a municipality that do not pay taxes, or a tax-paying entity such as a hotel, motel, or industrial plant; the savings will accrue from reduced electrical usage and the possible reduction in fuel costs due to the fuel user offering a cogeneration rate on the fuel, or a negotiated contract between the using facility and the fuel provider.

Tax-paying entities may also show additional savings due to being able to depreciate the equipment and deduct the maintenance costs as expenses in operating the cogeneration plant. Since each facility has its own peculiar need with regard to taxes and deductions, these items will be ignored in showing the overall savings a cogeneration plant has demonstrated to the client.

There are two ways in which those savings can be demonstrated:

1) Showing actual production of electrical energy and thermal energy and calculating the cost of such produced energy had it been billed to the client.

2) Comparing the overall cost of electrical and fuel bills from one period to another, before and after cogeneration.

While the first method is more scientific because it is pure in its statistics of produced energy, most clients will compare what they paid last year, before the cogeneration plant was installed, to what they paid this year, after the cogeneration plant was installed. They will then set their operating budgets for subsequent years reflecting those savings and be satisfied when actual expenses meet or better projected expenses. Obviously, the pitfall in this latter method is encountered when a facility has expanded and, with that expansion, more energy will be required and used. However, the utility still supplies monthly figures showing the amount of kilowatts the cogenerator has put out, and this is used as a check list to see if the system is performing as expected.

Therefore, the following case histories will show what the client used and paid for in the way of utilities over a twelve-month period before cogeneration was installed versus what was used and paid for in the twelve months after cogeneration was installed. None of these facilities underwent any significant expansion, so the figures are a good representation of the savings the cogeneration plant derived for the clients.

CASE HISTORY #1

A two-hundred room hotel installed a 40-kW cogeneration system. The thermal output was used to heat the water for the guest rooms, the laundry facility, and the swimming pool and spa. While the same heat output was used to heat all the water, separate heat exchangers were used for the potable water, i.e. the guest room and laundry water, and the swimming pool and spa water. Code requirements dictated that the potable water heat exchangers were designed with "double wall shell and tube" features so that if there were any tube leakage of ethylene glycol, the cogenerator engine's cooling fluid, it would not contaminate the potable water supply.

The initial investment of $75,000 was financed at 8.0% interest for a period of 5 years. The monthly payment was $1,511 versus a monthly savings of $2,612. This resulted in a positive cash flow of $1,101 per month *during the finance period.*

CASE HISTORY #2

A YMCA installed a 42-kilowatt cogeneration system to provide electricity throughout its system and to heat water for its swimming pool. Since no potable water was being heated, a plate-seal type heat exchanger was used exclusively to heat the swimming pool water.

The initial investment of $67,302 was financed over a period of 5 years. The monthly payment was $1,295 versus a monthly savings of $1,857. This resulted in a positive cash flow of $6,748 per year *during the finance period.*

CASE HISTORY #1

	BEFORE COGENERATION	AFTER COGENERATION
Electrical Usage:	731,220 kilowatts	428,820 kilowatts
Electrical Demand:	1,920 kilowatts	1,416 kilowatts
Gas Usage:	29,960 therms	5,832 therms
Cogen Gas Usage:	0 therms	42,624 therms
Electric Costs:		
Energy:	$76,778.00	$45,026.00
Demand:	$14,784.00	$10,011.00
Gas Cost:	$17,976.00	$17,749.00
Maintenance Cost:	-0-	$5,400.00
Total Costs:	$109,538.00	$78,187.00
Savings Per Year:		$31,351.00
Return on Investment:		39.2%
Payback, (Return of Capital):		2.55 years.

CASE HISTORY #2

	BEFORE COGENERATION	AFTER COGENERATION
Electric Usage:	258,480 kilowatts	88,028 kilowatts
Gas Usage:	32,366 therms	48,264 therms
Electric Cost:	$27,137.00	$8,763.00
Maintenance Cost:	-0-	$6,300.00
Total Costs:	$46,460.00	$24,174.00
Savings Per Year:		$22,288.00
Return on Investment:		33.0%
Payback, (Return of Capital):		3.0 years.

CASE HISTORY #3

A health and fitness facility installed a 60 Kilowatt cogeneration system using the electrical output throughout its facility and the thermal output to heat its shower water, swimming pool and spa.

	BEFORE COGENERATION	AFTER COGENERATION
Electric Usage:	1,009,058 kilowatts	654,069 kilowatts
Electric Demand:	2,333 kilowatts	1,814 kilowatts
Gas Usage:	37,991 therms	61,001 therms
Electric Cost:		
Energy:	$50,453.00	$32,721.00
Demand:	$43,160.00	$32,164.00
Gas Cost:	$27,236.00	$19,451.00
Maintenance Cost:	-0-	$8,900.00
Total Costs:	$120,849.00	$93,236.00
Total Savings:		$27,613.00
Return on Investment:		32.49%
Payback (Return of Capital):		3.08 years.

The initial investment was $85,000. The facility used internal financing to charge the project 6% interest over a 5-year period. The monthly payment was $1,635 vs. a monthly savings of $2,301 resulting in a positive cash flow of $7,993 per year during the finance period.

CASE HISTORY #4

In 1991, a large hospital in upstate New York installed five, 75-kW cogeneration units. In 1993, the system reduced the hospital's electric bills by more than $166,000, used less than $29,000 worth of natural gas and yielded a net energy cost savings of nearly $138,000. After maintenance costs of about $39,000, the hospital is saving over $99,000 per year.

The total run hours for the five units was 27,862 hours, or about 5,572 hours per unit. That factors out to a 71% availability, but since the system is designed to operate fully when the outside temperature is below 50°F, not all units are run during the summer time. Wintertime figures show about 96% availability with all five systems in operation, an enviable record.

The hospital is said to be contemplating putting in absorber-chiller air-conditioning units in the future allowing the cogeneration system to run longer hours during the summer thereby increasing the plant utilization and returning even more money in the way of savings to the hospital.

Over 2,144,000 kilowatts of electricity were generated during the year, and it was estimated that 1,910 kilowatts in electric demand were saved. The combined energy and demand cost savings showed that the electric utility was billing the hospital at a combined rate of $0.0777 per kilowatt.

The cost of the system was $1735 per installed kilowatt, or about $650,000. The state granted the facility a $250,000 no recourse grant leaving the hospital with a $400,000 capital outlay. The $99,000 annual savings, therefore, shows the return on investment to the hospital to be about 25% or a 4-year payback. One item in the balance sheet showed that the net increase in natural gas usage of about 72,000 therms resulted in a net gas cost increase of almost $29,000. If this facility had the benefit of a "utility gas rate" schedule, the gas costs may have shown a net decrease in the cost of gas, leading to an even greater return on investment.

CASE HISTORY # 5

A manufacturer of food products had a need for a constant supply of chilled water to keep his cold rooms cool. Forty-three-degree Fahrenheit chilled water was needed to accomplish this. He decided to put in 232 kWh of on site generation with half of the generated heat going to a 42-ton absorber chiller that was direct fired from the exhaust of the turbines. The other half of the produced waste heat went to feed an air to water heat exchanger to produce 180°F hot water.

The results were as shown in the chart on the following page.

Case histories and customer references are the lifeblood of the small-scale cogenerator manufacturers' success. To insure that success the machines must operate, and to operate they need the monitoring and

Case History Showing Effects of 42 Tons of Absorber Chiller

	Before Cogeneration	After Cogeneration
Electrical Usage	2,360,000 kWh	388,000
Displaced kWh	0	336,000
Net kWh Consumption	52,000	
Electrical Demand	4,396 kW	1,612 kW
Displaced kW	0	640 kW
Net Demand Consumption	972 kW	
Gas Usage	64,880 Therms	22,800 Therms
Fuel Usage	0	275,740 Therms
Electrical Costs:		
Energy:	$266,680	$5,876
Demand:	$33,102	$7,320
Gas Cost	$22,708	S%,509
Maintenance Cost	0	$12,000
Total Costs	$322,518	$121,705
Depreciation Tax Savings	0	$21,163
Savings per Year	$221,976	
Project Cost	$503,883	
Return on Investment	44.05%	
Simple Payback in Years	2.27 Years	

NOTE: The absorber chiller uses very little electricity to produce the 42 tons of chilled water. Conventional electric-driven chillers use anywhere from 1.2 to 2.5 kWh/ton of produced chilled water, or over 428,000 kWh per year t the 1.2 kWh/ton figure.

maintenance supervision that most manufacturers offer. Very few commercial clients have the available expertise in personnel to keep an on site cogeneration plant functioning well. Even personnel who may have good automotive engine skills do not realize how important the cooling system is to successful cogeneration. If the engine heat is not removed from the coolant in a well-balanced thermal-recovery system, the engine will overheat and shut down or, worse yet, run too cool and cause condensation in the oil which then adversely affects all the internal combustion parts of the engine.

Case Histories

This author has first-hand knowledge of over one hundred small-scale cogeneration systems. Only one installation did not include the manufacturer's maintenance program, by choice of the client, and that installation fared very poorly in on-line availability. Similarly, when a cogenerator manufacturer ceases business, it has been very difficult for that installation to operate as successfully as it had under the manufacturer's care and maintenance.

The following chapter will discuss the current manufacturers of small-scale cogeneration as well as some of the past manufacturers and how the industry in general must collaborate with one another when one manufacturer no longer has the wherewithal to maintain its installed cogeneration systems.

Chapter 13

Small-scale Cogeneration Manufacturers

In 1995, the author was commissioned by Hitachi Building Equipment Engineering Company to do a study on the manufacturers of small-scale cogeneration. Ten companies were identified as being manufacturers, assemblers, or developers in the cogeneration field. Other than the two developers, of the eight manufacturers or assemblers identified in 1995, it appears only four viable companies remain in the field of manufacturing small-scale cogeneration units.

This author has been on or near the cutting edge of the evolution of small-scale cogeneration due to the early developments that occurred in Southern California. During the 1980's, a small company called Thermex manufactured a 10-kilowatt cogeneration unit. It used a Ford engine with a natural gas carburetor driving a generator that was manufactured by Baldor. A coiled heat exchanger acted as the interface between engine coolant and the water to be heated, while an exhaust-gas exchanger wrung the last ten percent of the engine's heat from the exhaust. Very little has changed in this rudimentary explanation of cogenerator design since that time.

Three enterprising people saw merit in what Thermex was doing and became distributors of this product. They were Herbert Ratch, Margo Ratch and Dave Lumbert. Their expertise was in sales and marketing. Soon, they were selling these ten-kilowatt units almost as fast as they were being manufactured. Coin-operated laundries were the prime target and what better industry to target. Each laundry used abundant quantities of hot water and electricity and were small enough to support a 10-kilowatt unit, while paying the highest commercial rate for electricity—virtually a residential rate. The units sold for approximately $10,000 installed and had simple paybacks of less than two years. A fifty-percent-or-better return on investment!

It seemed to the manufacturer that these three entrepreneurs were dictating where the marketplace was going, and due to fear of not controlling its own destiny, started to limit the success they were having. Undaunted, the Ratch's and Lumbert decided to design their own unit, and under the corporate name of Cogeneration Energy Services, built their first 10-kilowatt units out of their office suite in Newport Beach, CA. It was at that time that this author became acquainted with Herbert Ratch, who was also doing consulting work in the waste-to-energy field, an area in which this author was also working for a waste to energy boiler manufacturer.

In 1987 the Ratch's convinced this author to become part of their operation as an independent distributor of their cogeneration units. Herb Ratch and I traveled to Atlanta, Georgia to attend the Coin Laundry Expo, a national convention of suppliers to that business. With nothing more than a table and some literature, the booth that we manned at this Expo was inundated with curiosity seekers wanting to learn about this magic box called cogeneration. Cogeneration Energy Services came away with over three hundred leads for cogeneration sales and over one hundred interested people to become distributors. This was exciting, but limited capital resources prevented exploiting these leads fully, and the company remained pretty much a California-based entity serving the needs they found locally. But, after all, California is the world's 6th largest economy, so concentrating there was not all that bad.

Meanwhile, designs for a 20-kilowatt unit were in place, and Cogeneration Energy Services changed its name to North American Cogeneration Company. A third-party assembler was found, and units were manufactured in a factory atmosphere. As sales for North American Cogeneration Co.'s 10- and 20-kilowatt units grew, Thermex was finding it harder to keep pace and soon they ceased operations. This led to a public-relations dilemma. Should the users of Thermex cogenerators be allowed to flounder and find their own service outlets, or should North American Cogeneration step in and be the stepfather to these orphans, many of whom were sold by the antecedent of North American, but not serviced by them? A program was instituted to take over the service of these units thereby keeping the image of on-site cogeneration viable for new sales in established markets. Through the auspices of the cogeneration associations that were forming in California, this program was formalized and any user of the Thermex units was able to sign on with North American Cogeneration for regular servicing of their units.

As North American Cogeneration Company was becoming established in the 10- and 20-kilowatt-size units, a company that has been a leader in small-scale cogeneration for many years, Tecogen, was marketing 30-kilowatt units and larger. Tecogen is a division of Thermo Electron Company out of Waltham, Massachusetts, and is still very much a player in small-scale cogeneration. However, their success has been primarily in the eastern part of the country, and relatively few units were sold and installed in the West. Today, Tecogen specializes in 60- and 75-kilowatt, packaged, cogenerators.

In the late 1980's, a man named Craig Linden developed a 22-kilowatt unit under the name of Microcogen and commenced marketing that product. His was a very compact unit that had clean lines and good technology, but the very compactness that produced some of those attributes made it a maintenance nightmare. Microcogen formed a relationship with Ultra Systems, a highly successful defense contractor out of the Los Angeles area, which undertook the manufacture of the Microcogen units. Their star rose quickly, but fell almost as quickly mostly due to the maintenance shortcomings.

It was becoming obvious to North American Cogeneration that good competition in this specialized field was needed in order to cause the industry to flourish. It reminded the author of Abraham Lincoln's phrase that he was a starving lawyer in Springfield, Missouri until a second lawyer moved into town. The same was needed in small-scale cogeneration so its attributes could be measured competitively and not in an exclusive market. Microcogen could have supplied that impetus, especially with the credibility of Ultra Systems behind them.

There were some other small-scale manufacturers around, but they seemed to specialize in niche markets. A belt-driven unit manufactured in Arizona sold mainly to Japanese markets. Caterpillar and Waukesha, two sound engine manufacturers entered the field with smaller units but did not have the packaged concept. They were mainly skid mounted engine-generators having no acoustic packaging and consequently required sound proof rooms or remote locations in which to be installed. Unattenuated engines would reach decibel readings of 90 or more and were not suitable for commercial establishments surrounding residential neighborhoods and certainly not tenable in hospital, nursing home, hotel, and restaurant venues. The packaged units, such as North American's and Microcogen's were enclosed and could boast decibel readings of 70 or less within 6 feet of their installations.

In the early 1990's, a manufacturer came into being under the name of Intelligen Products. They offered cogeneration packages in the 10, 20,

40, and 60 kilowatt sizes. They took many of the attributes of the existing units such as water-cooled manifolds and expanded metal exhaust gas heat exchangers, but incorporated many design features that made them more reliable. Directly connected engine-generator design eliminated the troublesome coupling; expanded oil reservoir capability eliminated the need for frequent site visits; sophisticated control technology with built in remote monitoring allowed knowledge of system operation and expanded routine monthly servicing to every 2000 hours of run time. Attractive packaging made the units more aesthetically pleasing with easier internal access for maintenance without destroying the compactness features necessary for many commercial sites. This was the competitor that this still fledgling industry needed, and trade shows now offered choices to the potential client.

Then John Hanna stepped into the picture. North American Cogeneration realized its need for better capitalization especially now with the advent of a viable and attractive competitor like Intelligen Products. Herb Ratch met John Hanna during a trip to Houston in 1992, and introduced Hanna to the world of small-scale cogeneration. Hanna was running a company that specialized in financing energy-related projects, often on a shared-savings basis.

Cogeneration was a natural for this type of contract, and North American had already installed a system using third-party financing under a shared-savings contract at La Quinta Inns in Costa Mesa, California. C. Itoh, the Japanese conglomerate, acted as the financier.

Hanna showed interest, and one thing led to another with Hanna coming to California and structuring a deal whereby he would capitalize North American Cogeneration in return for a controlling interest in the company. Ratch and Hanna formed an affinity with one another that did not extend to Dave Lumbert and Margo Ratch. Hanna and Lumbert, especially, were like fire and water. Hanna decided he had to have Lumbert out of the business of North American Cogeneration and tried to demean him with meager assignments around the plant such as inventory taking and parts procurement. Unfortunately for Hanna, it was Lumbert who knew more about the day-to-day operation of the business than any other single person in the company. He knew how to manufacture the units, how to install them, and how to service them. Without those three key ingredients the business would falter. Lumbert also held another ace up his sleeve, that of the transfer of his stock to Hanna to make control complete for Hanna. Since each of the three partners held equal interest in the business, Hanna needed at least two of them to sign over their stock for him to have majority control. Margo Ratch was more aligned with Dave Lumbert than with her own husband and refused to sign over her share of the company until Hanna made good on his promises of dealing with the creditors: putting the agreed-upon monetary exchange in escrow and meeting other stipulations of the buy-out agreement. For some reason Hanna was moving very slowly in those areas.

Hanna had, in the meantime, offered this author a consulting position in the company to take what had been built as a successful distributorship in Southern California throughout the state. A monthly stipend was agreed upon and not knowing the undercurrent that was prevailing in the company, I proceeded to seek and find quality distributors in northern California and other parts of the state. In a private conversation with Hanna during a luncheon meeting he advised me he was considering bankrupting North American Cogeneration and restarting the business under another name just to rid himself of Lumbert and Margo Ratch. He promised that Herb Ratch, he (Hanna) and I would be the integral parts of the new company. By this time, I had evaluated Hanna's dealings as somewhat spurious, cautioned him against taking this action, and reminded him that we were about to sign on a quality distributor in the San Francisco area that was willing to place an order for at least six units for stock. Upsetting the company by taking such drastic action simply to remove Lumbert was foolhardy.

Hanna took the company into bankruptcy the following week. In order for such a bankruptcy to take place, the presiding judge required that the books of the company be produced to verify creditors, assets, liabilities, etc. Hanna could not produce the books as Lumbert had adroitly spirited them away in anticipation of such action. This led to the bankruptcy proceeding to be thrown out of court and with no corporate documentation, Hanna was made liable for debts and ongoing liabilities of the company.

Meanwhile, Hanna had written me checks for $5,000 which had bounced. Upon notifying him of such he wrote additional checks that also were unable to be cashed due to insufficient funds. He refused to make good on what he owed, so I filed a small claims action against him and was awarded a $3,000 judgment against him in December of 1993. Hanna appealed this judgment, and the appeal hearing was scheduled to take place in Riverside, California, on February 8, 1994.

On February 7, 1994, I was in my office, which was in my home in

Carlsbad, CA, doing, among other things, preparing for the next day's appeal hearing against Hanna. About noon the phone rang and it was the Santa Ana, California, police department calling advising me that a colleague of John Hanna's, Greg Lawyer, had been shot and severely wounded by Hanna inside the parking garage adjacent to Lawyer's office. The police knew nothing more at that time except that Lawyer, having been shot twice, one bullet just missing his heart, had the presence of mind to ask the police to seek the address book in his office desk and call the people in that book advising them of this shooting. I was in Greg Lawyer's book, hence the call.

Musing over this information left me bewildered, but somewhat cautious so I affixed a small padlock to the gate outside my house and proceeded to call the Riverside court to advise them of what I had learned and to inform them I would not be appearing in court the next day. Around 2:00 p.m., the phone rang and it was the bailiff at the Riverside court calling to find out more about John Hanna as they had heard about the shooting and wanted a detailed description of Hanna in case he should show up the next day at the appeal hearing.

As I was speaking to them I heard a shot ring out, and my reflexes immediately propelled me to the floor of my office with the phone still at my ear. I told the bailiff what I'd heard and then, looking through the slatted blinds of my office window, saw the head and shoulders of John Hanna peering into my window! Hanna had shot the small lock off the gate and had gained entrance into my yard. The bailiff suggested hanging up and dialing 911, which I did in an instant. As I was informing the Carlsbad police of the situation, the doorbell rang sending a cold chill up my spine. I assumed the door was locked, but could not proceed to any other part of the house without being viewed through the many windows and sliding glass doors that prevailed in the house. The garage was my only refuge and there I went.

Ensconced in a makeshift closet in the garage, I armed myself with a stout table leg and had the advantage of being in the dark with the door to the garage backlighted should Hanna gain access to the interior and decide to investigate the garage. Moments later I heard another shot ring out and figured Hanna now had gained access to the interior by shooting the lock to the front door.

Then, I heard voices outside the garage. Not knowing who or what they might be, I remembered my cell phone in the car and quietly opened the car door and retrieved the phone and called 911 again. I was patched in to the police cruiser that was now in front of my house and was advised to open the garage door and come out with my hands up as the police, of course, did not know who I was. I obeyed and was ever so glad to see the police cruisers and accompanying police men outside my garage.

Not knowing if Hanna was in the house, the police dispatched a dog to go in and either flush him out or be sacrificed in the doing. The dog entered the house and after a short interval was followed by his armed handler only to find the house empty. Hanna had escaped by going over the wall adjacent to the street in back of the house. The shot I'd heard was an exasperation shot taken at the sliding glass door to my bedroom. The hole through the glass was clean and the bullet was imbedded in the Berber rug at such an angle that it was figured the shot was fired as Hanna was part way over the wall. A moment later the damaged glass in the door came crashing down in a shower of pebbles causing me and the two officers in my bedroom to flinch noticeably.

Hanna was on the loose and I needed no further encouragement when the police officer suggested I might be safer elsewhere until Hanna was found. I called a friend and took refuge in his house and then learned of the chain of events that led up to what transpired at my house that afternoon.

Hanna had made a breakfast meeting date with Herb Ratch at a Denny's restaurant somewhere in Costa Mesa. He told his wife, Irene, that he was going to "get them all" according to accounts of a phone conversation Hanna had had with a newspaper reporter later that day. She was reported to have said, "Go get them, John. Get them all!" and then he shot and killed his wife.

Ratch, arriving at the assigned restaurant with his new wife, Jeannette, sat and had coffee with Hanna. Jeannette then left to do errands, and Ratch and Hanna left to purportedly visit a contractor in Hanna's car. While in the car, Hanna shot and killed Herb Ratch and proceeded to dump his body near some grain silos in Chino. He then proceeded to Greg Lawyer's office. Lawyer was an investor in North American Cogeneration and a minor stockholder as well as a member of the board of directors of North American. His dealings with Hanna were supposedly cordial even though he now had lost his investment in North American Cogeneration. Hanna, according to Lawyer, came to his office unannounced and asked if they could go to lunch. Lawyer agreed provided it could be done quickly. Upon entering the attached parking garage and proceeding to Hanna's car, Lawyer went around to the passenger side only to be followed by Hanna who was pointing a gun at Greg. Surprised, but cogent, Greg lunged at Hanna as Hanna fired twice. One bullet narrowly missed Lawyer's heart exiting out his back, while the other bullet pierced his arm, also exiting. Hanna now noticed other people coming into the garage and quickly drove away leaving Lawyer for dead.

Miraculously, Lawyer not only survived but was immediately aided by the people who came into the garage to a trauma center also located within the office complex. While being taken to the trauma center he informed the people of the address book in his office and the request to call those people informing them of Hanna's action. Had it not been for the luck of Lawyer's survival and quick thinking, I would have opened the door to Hanna's doorbell ring and this story would have been written in an obituary column.

While taking refuge at my friend's house for the balance of February 7 and up until 4:00 p.m. on February 8, I was in touch with various police departments and newspapers to learn of any news and give them any assistance I could regarding Hanna's possible whereabouts. It was reassuring that Hanna had no knowledge of my friend and my car was placed in my friend's garage to keep it out of sight. Yet, it was a very unsettling 24 hours spent in hiding. About 4:00 p.m. on Tuesday, February 8, I received a call from the police department in Upland, California. They informed me that Hanna had returned to his apartment in Upland that afternoon. The police had staked out his apartment, and when they saw him, they approached his car cautiously. Hanna, upon seeing them, took the revolver he had used so ghastly the day before and ended his own life with one shot. The ordeal was over with the best possible conclusion. Had he not been captured what would I and the people who he had not yet "gotten," such as Lumbert and Margo Ratch, have done for the ensuing time? Even in capture, a trial would be held and facing him, not knowing what cohorts he might have engaged in his grizzly enterprise, would create a misery of its own.

In the aftermath we learned that John Hanna had a brother who is a doctor in Toronto, Canada. He verified that John was destitute. Any bravado he displayed as a financier was with other people's money and that he, the brother, had loaned John money just recently to meet everyday expenses. John Hanna was facing personal bankruptcy and ruin and could not fathom that mark against his heritage and decided to go on the rampage he did with the outcome only different by the number of people he would have "gotten."

Shortly before Hanna took North American Cogeneration into bankruptcy, I had been approached by Intelligen Products to represent them in Southern California, and so I did. The now "orphaned" users of North American Cogeneration had to be serviced in order for the industry to not be spoiled, and that was accomplished with varying degrees of success. Intelligen Products now offered the latest in the evolution of small-scale cogeneration with a more sophisticated, better designed and more reliable unit, but it was not without its own intrigue.

Intelligen Products' units were assembled in San Marcos, California, under the tutelage of a company called Coast Container Corporation owned by its president, Mr. Ray Raffesberger. That arrangement was tenable until the owners of Intelligen decided they wanted to break away from that arrangement and did so by changing the name of the company to Intelligent Solutions, Inc. (ISI) and striking a deal with MagneTek Corporation, a Fortune 500 company and manufacturer of engine-generator sets, to be the assembler of Intelligent Solutions cogenerators.

In so doing, ISI left in its wake the remnants of Intelligen Products, and Mr. Raffesberger picked up those remnants and turned it into Coast Intelligen Corporation, a manufacturer of packaged cogenerators in the 35- and 60-kilowatt sizes.

Thus, from six or eight manufacturers of small-scale packaged cogenerators in 1995, the new millennium will see three or four quality manufacturers that have weathered the storms and may be able to say the industry has arrived as a viable supplier of a technology, which when applied properly, offers all the promises originally made: Economy, Efficiency, Ecology.

New manufacturers of small-scale cogeneration systems have emerged in both packaged reciprocating engine driven units as well as microturbines. As mentioned in other parts of this book, small-scale cogeneration is normally believed to fall in the category of 0-500 kilowatts. While companies such as Caterpillar, Waukesha, Cummins and other engine manufacturers certainly have built and will continue to build cogeneration systems, they are considered "assembled" units as opposed to "packaged" systems. The same holds true for gas turbine manufacturers such as Solar and General Electric.

Therefore, the following listing is categorized by first, Reciprocating Engine Driven Packaged Cogeneration systems and Microturbine Driven Cogeneration systems. All fall into the less than 500 kW size range.

RECIPROCATING ENGINE DRIVEN
PACKAGED COGENERATION MANUFACTURERS:

- Tecogen, 45 First Avenue, Waltham, MA 02451 Phone: 781-466-6400

- Coast Intelligen, Inc., 2460 Ash Street, Vista, CA 92083 Phone: 760-597-9090

- Hess Microgen, 12 Industrial ParkWay, Unit B I, Carson City, NV 89706, Phone: 775-884-1000

- Trigen Energy Corporation, 1177 West Loop South, Houston, TX 77027, Phone 713-552-2039

- Bluepoint Energy Products, 10 E. Greg Street, Sparks, NV 89431, Phone: 775-786-1332

The packaged reciprocating engine sizes range from a low of 60 kWh to a high of over 400 kWh, but most of these manufacturers are preferring to offer the 250 kWh as their flagship size.

MICROTURBINE COGENERATION SYSTEMS

- Capstone Turbine Corp., 21211 Nordhoff Street, Chatsworth, CA 91311 Phone: 877-716-2929

- Ingersoll-Rand Power Works, 30 New Hampshire Ave., Portsmouth, NH 03081 Phone 877-IR POWER

- Turbec Americas, Inc., 215 Celebration Place, Ste. 500, Celebration, FL 34747 Phone: 321-559-1005

- Bowman Power Systems, Inc., 20501 Ventura Blvd, Ste. 285, Woodland Hills, CA 91364 Phone: 818-999-6709

The Microturbine systems range in size from 3 0 kWh to 100 kWh with larger sizes predicted to be on line as large as 200 kWh by 2003 or so.

Any manufacturers who have been left out of this list have been done so without any purpose other than attempts to contact the original 1995 manufacturers have met with silence or returned correspondence, and any new manufacturers were not known to the above-listed ones when asked about their competition.

Chapter 14

Do-it-yourself Cogeneration

Anyone can determine if cogeneration is viable for a given facility simply by using the information readily available from the monthly utility bills and a few tips that are given in this chapter.

First, it must be determined that there is a need for the products of cogeneration. Electricity is almost a given as it would be hard to envision any facility that does not use electricity. The other product, i.e. hot water, steam, or air conditioning, requires some additional analysis.

If your facility uses hot water as part of its everyday business then this would be the most likely "second" work effort to analyze to see whether cogeneration is economically feasible for the facility. The key words here are "economically feasible." Regardless of how idealistic one may be to want to conserve fuel and aid the environment, a project must have an economic benefit before it will be implemented.

Look at your hot-water-usage points. If you are a hotel, you will use hot water in your guest rooms, kitchen, laundry, swimming pool and spa. If you are an industry, look for hot-water usage in your product-manufacturing area. Is metal being plated in hot water or hot chemical baths? Is food being cooked? Is hot water used for constant cleaning of machinery? If you are a health-and-fitness facility, look for the heating of water for showers, swimming pools and spas. If you are a restaurant, your kitchen needs for dish washing. A municipality uses heat for the swimming pools, showers and spas. If you are an office building, there is probably very little hot water used directly, but air conditioning may be a way to use hot water in an absorber-chiller unit to supplement your electric air-conditioning units. This latter is a special application and will be discussed after the more normal uses of hot water are analyzed.

If your facility uses no hot water, steam, or air conditioning; you are probably not a good candidate for the benefits of cogeneration and your analysis will stop there.

COLLECT THE DATA

Assemble your utility bills for the last year. Then, using a form such as shown in Figure 14-1, compile the usage of electricity and fuel on a month-to-month basis. Input both the quantity of electricity in kilowatts as well as the cost. If your electric bill is broken down in both energy costs and demand costs, then list it in that manner. Your fuel costs should also be listed in quantity and cost for the energy supplied. This is often shown as therms in the case of natural-gas usage. If fuel oil is used, this will be shown in gallons delivered periodically so the total delivery for a year should be added up to determine how much fuel oil is used annually. The table in Appendix II will show the conversion factor for the heating energy in fuel oil expressed in therms. The same is true for auxiliary types of fuel such as propane where the delivery is periodic and expressed in pounds or gallons of propane. Again, the conversion factors shown in Appendix II would be used to convert that usage to therms.

Next, fill out the top portion of the form. Obviously, the first four lines are simply describing who you are. Starting with Electric Utility and Fuel Utility, show the supplier of both these commodities. Skip the Thermal Load Fraction for the moment. Indicate the size of your hot-water-storage tank(s), the hours your facility is open, and the voltage and phase of your electric supply. They are found on your main electric distribution panel. Usually, most commercial and small industrial facilities would use 240 or 480 volts and are 3-phase, 60-hertz electric systems.

Heating-system efficiency is a reflection of your hot-water heater's output in Btu's vs. input of fuel in Btu's. If the heater shows an input of 360,000 Btu's of natural gas and an output of 300,000 Btu's of hot water, its efficiency is 300,000/360,000 or 83% efficiency. Most hot-water heaters have an efficiency rating between 70 and 85%. The older the heater, the less efficient it may be due to dirt accumulation or plating of the heating elements by the minerals in supplied water. For every 5 years your heater has been in service, it would be prudent to derate the efficiency by 3 points.

Thermal Load Fraction is simply the amount of natural gas (fuel) used versus that used to heat water.

Space heating is the most common "other" use of supplied fuel. Therefore, look at your winter usage of fuel versus your summer, spring and fall usage. Two good months to compare are February and May. February is usually the coldest month while May is warm enough that

Table 14-1. Cogeneration Survey Form

DATE OF SURVEY: _____

NAME OF COMPANY: _____ CONTACT: _____

ADDRESS: _____

PHONE: _____ FAX: _____

ELECTRIC UTILITY: _____ FUEL UTILITY: _____

THERMAL LOAD FRACTION: _____

STORAGE TANK SIZE: _____

HOURS OF OPERATION: _____ ELECTRIC PHASE: 1 () 3 ()

VOLTS: 208/240 () 460/480 () HEATING SYSTEM EFFICIENCY: _____

MONTH	FUEL USAGE		ELECTRIC USAGE			
	THERMS	$	ENERGY (kW)	$	DEMAND (kW)	$
JANUARY						
FEBRUARY						
MARCH						
APRIL						
MAY						
JUNE						
JULY						
AUGUST						
SEPTEMBER						
OCTOBER						
NOVEMBER						
DECEMBER						
TOTALS						

space heating is at a minimum while other hot-water usages are relatively constant. The difference in usage between those two months is a good representation of fuel that is used for non-hot-water heating.

Other usages for fuel, especially natural gas, may be cooking and air drying of laundry. Look at the nameplate reading of your stoves and dryers to determine how much fuel they consume and then estimate how many hours per month those fuel using devices are used. Subtract those usages from your total fuel usage as well as that used for space heating.

Example: Suppose your February fuel bill is 10,000 therms and your May fuel bill is 6,000 therms. It's safe to assume that most of those 4,000 therms were used for space heating. If you are also heating a swimming pool, that requires much more heat in the winter time than during other seasons. By using the charts in Appendix III, you can determine the overall thermal needs required to heat swimming pool water to 78 - 82 degrees. Therefore, you can fine tune your winter space-heating requirements by seeing how much fuel your pool required during the winter period. If a spa is in use, it's fuel consumption is pretty constant throughout the year since most spas are heated to 104 degrees all the time, and the difference in outside temperature is a small correction for that relatively small body of water.

From those observances you calculate your overall space heating and other fuel usages to be 30% of your overall annual fuel usage. Subtract 30% from 100% and your Thermal Load Fraction is 70%. That is the fuel you will try to displace with cogeneration.

Table 14-2 shows an example of a filled in Cogeneration Survey Form. Inspection of the data shows more fuel used during the winter than in the summer but more electricity used in the summer than in the winter. The reason for both is climatization of the facility: heat in the winter and air-conditioning load during the summer. The swing months during spring and fall show usages that minimize the climatization factor.

The data sheet shows that the facility is open, or at least using energy, 7800 hours per year. Since there are only 8,760 hours in a year, we may consider this facility to be open virtually 100% of the time.

Dividing the total kilowatts used: 456,998 by the number of hours the facility is open: 7800, a figure of 58.59 kilowatts per hour is derived. If we know from vendor data that a 60-kW cogeneration unit is manufactured, that may be the initial selection. Seventy-eight hundred hours a year is equal to 650 hours per month. Multiplying 650 hours per month

Table 14-2. Cogeneration Survey Form

DATE OF SURVEY: AUGUST 9, 1999

NAME OF COMPANY: ABC FITNESS CENTER CONTACT: JOHN SMITH

ADDRESS: 2500 ELM STREET, ANYTOWN, USA, 92000

PHONE: 213-555-4000 FAX: 213-555-5000

ELECTRIC UTILITY: CONSTANT ENERGY CO. FUEL UTILITY: GAS SUPPLY, INC.

THERMAL LOAD FRACTION: 70% STORAGE TANK SIZE: 500 GALLONS

HOURS OF OPERATION: 7800 HOURS/YR. ELECTRIC PHASE: 1 () 3 (×)

VOLTS: 208/240 () 460/480 (×) HEATING SYSTEM EFFICIENCY: 70%

MONTH	FUEL USAGE THERMS	$	ELECTRIC USAGE ENERGY (kW)	$	DEMAND (kW)	$
JANUARY	13,794	$7,040	38,920	$2,335	145	$1,305
FEBRUARY	8,159	4,683	33,320	1,999	135	1,215
MARCH	9,108	5,116	37,520	2,251	143	1,287
APRIL	8,027	4,440	38,720	2,323	145	1,305
MAY	5,413	3,050	34,958	2,447	137	2,055
JUNE	4,459	2,645	33,320	2,332	134	2,814
JULY	1,947	1,237	43,720	3,060	160	3,360
AUGUST	489	310	41,600	2,912	155	3,255
SEPT.	1,232	774	45,280	3,169	165	3,465
OCT.	2,072	1,297	38,320	2,399	144	1,296
NOV.	5,643	2,998	37,880	2,273	143	1,287
DEC.	8,147	4,202	33,440	2,006	135	1,215
TOTALS	68,517	$35,411	456,998 kW	$29,506	1,741	$23,859

COST/THERM: $0.517   COST/kW: $0.0646   COST/kW DEMAND: $13.70

times 60-kW/hour gives us a figure of 39,000 kW per month. If we applied that size cogenerator to the facility it would overproduce electricity in all but the three summer months of July, August, and September. That means electricity would be overproduced during the other nine months of the year.

We can also derive that since the demand usage is well above the 60-kW preliminary selection, most of the electricity is used during the daytime when the facility is at its peak operation. The average demand is 145 kW per month.

That analysis tells us that the facility uses most of its electricity during the daytime, and our 60-kW cogenerator would overproduce electricity by a considerable amount during the off-demand nighttime hours. We may consider a smaller-size unit, or we might consider the 60-kW unit, but only run it 18 hours per day instead of the 24.

Before we make a final decision, let's look at the thermal side of the equation. As a rule of thumb, for every kilowatt a cogenerator delivers, about 8,000 Btu's of thermal energy is delivered. Therefore, a 60-kW cogenerator will deliver about 480,000 Btu's of thermal energy, or 4.8 therms/hour. Operating 7800 hours per year, the cogenerator would deliver 37,440 therms of heat energy. This is well below the needed 68,517 therms the facility consumes. But the thermal load factor, i.e. the amount of energy the facility consumes to heat water is 70%. Therefore, the usable offset of thermal energy is 68,517 therms times 70% or 47,962 therms. Our delivered 37,440 therms is still below this amount on an annual basis, indicating an acceptable fit. Monthly, the unit would deliver 4.8 therms times 650 hours, or 3,120 therms per month. This is acceptable in all but the months of July, August, September and October. During those months thermal energy would have to be dissipated via a blow off radiator.

Will the selection of a 60-kW cogenerator derive an economic benefit to our facility?

ECONOMIC ANALYSIS

Another rule of thumb is the installed cost of a small-scale cogeneration system is about $1500/kilowatt. Therefore, our 60-kilowatt system would cost $90,000 installed. Add to this operating costs for fuel and maintenance. Will this system give us a desired return on investment?

Electric Offset

Generate 468,000 kilowatts of energy @ $0.0646/kilowatt = $30,233 per year savings. Sixty kilowatts of connected load would displace that amount of demand cost. Therefore, 60 kilowatts times an average of $13.70 per demand kilowatt = $822 per month times 12 months = $9,864 in demand savings. Therefore, electric savings potential is $40,097 per year.

Gas (Fuel) Offset

Generate 37,800 usable therms of heat energy @ $0.517/therm = $19,543 annual savings.

Total Savings

$40,097 in electrical savings + $19,543 in gas savings = $59,640 total gross savings in combined electrical and fuel costs.

Operating Costs

How much gas will be used to run the cogenerator. Another rule of thumb is that for every kilowatt of produced electricity, approximately 13,000 Btu's of gas is required to run the unit. Therefore, 780,000 Btu's of fuel are required, or 7.8 therms per hour. If your gas company will give you a special rate based on constant usage you may be able to buy the gas for less than the commercial rate. The gas company will generally sell to the utility at a rate of $0.28 to $0.35 per therm. Therefore, at $0.35/therm, the cost to run the cogenerator is 7.8 therms per hour ($0.35 = $2.73 per hour (7800 hours = $21,294 per year.

Maintenance will run $8,970 (see Chapter 9), so total operating costs will equal $30,264 per year.

Net Savings

Generated savings = $59,640 per year versus operating costs of $30,264 realizes a net savings of $29,376 per year. Will this be incentive to go ahead with the cogeneration plant?

Simple Payback

Installed cost of $90,000 divided by net savings of $29,376 = 3.06 years simple payback or 32.64% return on investment. It appears a cogeneration project would be worth pursuing and a call to the manufacturer would be in order.

The manufacturer's representative would fine tune all these numbers, and by use of load meters, his recommendations would be based on the minimum overproduction of electricity, if any, and he would arrive at a selection that would be optimum to the facility's needs. By having gone through your own analysis, however, you would be able to insure what the manufacturer is telling you correlates with your own findings. Had your own analysis derived a payback that was marginal, you could share those figures with the manufacturer to see where his figures would differ with yours, but in all likelihood a marginal payback using your analysis would not be much improved upon. If anything, cost of fuel and downtime during peak periods would detract from your ultimate savings, and other pitfalls may come into play.

In conclusion, if your own figures can show a 25% return on investment or better, i.e. a 4-year payback, it would behoove you to contact a cogeneration system manufacturer to analyze your system thoroughly.

To help facilitate employing actual production numbers, Appendix IV shows some average production numbers derived from a composite of various manufacturer's specifications.

If your usage of hot water is marginal but a significant air-conditioning load is evident, cogeneration still may be viable by using absorber chillers to offset some of the electricity used by the conventional air conditioning units. Absorber chillers operate under the principle of boiling a fluid under very low pressure to produce a refrigerant vapor that is used to chill water. The chilled water circulates through the building where fans blow air over the chilled water pipes to produce cool air into the rooms to be cooled. The boiling of the refrigerant fluid can be accomplished by either a direct flame or by introducing hot water as the heat source. About 10 tons of cooling can be produced from every 30-kW cogeneration system.

Since absorber chillers produce air conditioning with little electrical usage, while conventional air conditioners use 1.2 kW for every ton they produce, the use of cogeneration for this purpose produces a savings in overall electrical costs. In the above example of a 60-kW cogeneration unit, approximately 20 tons of air conditioning can be produced saving 24 kilowatts of electricity every hour. The savings in energy cost is $1.56 per hour and $312 per month in demand costs. If the air-conditioning system operates 300 hours per month, the electrical savings would be $468 in energy costs plus $312 in demand costs for a total savings of $780 per month. Since modern buildings run a combination of heating and air conditioning year round, the annual savings may amount to over $9,000 per year.

Often, the application of absorber-chiller technology in conjunction with cogeneration is best accomplished when the facility is contemplating adding air conditioning to the facility. Thus the savings are not only in future operating costs, but the absorber-chiller capital expense replaces most of the cost that would have been used for the conventional electric air conditioning unit.

Chapter 17 discusses absorber-chiller technology in more detail.

Chapter 15

Green Energy vs. Cogeneration

Much is being heard about the utilization of "Green Energy." This reference is applied to the production of electricity from natural or renewable sources. Those natural sources are defined as anything that does not consume a resource that may run out, such as oil, coal or natural gas. Such sources are:

>Geothermal
>Biomass and waste
>Small hydroelectric
>Wind power
>Solar energy

Such plants are often small and remote, meaning that the electricity they produce often is more expensive than power generated by large, conventional plants that burn gas or oil or employ nuclear power.

The impetus in Green Energy is a result of the deregulation of electrical supply in those states that have allowed that sort of competition (in other words, where customers have the ability of choosing their own electrical supplier, not that electricity from these sources is brand new, but the fact that this electricity can now be sold to anyone whereas previously it had to be used on the site where it was generated, or sold to the prevailing utility at the utility's avoided cost).

In California, companies have been formed that not only offer conventional electricity direct to the consumer, but also offer Green Energy to that consumer. To date, the cost of Green Energy is advertised as being slightly higher per kilowatt to the consumer than conventional energy. The California Energy Commission is collecting a small surcharge from all Californian's power bills and, through a fund called the Renewable

Resource Trust Fund, is subsidizing renewable energy to keep the cost of this energy to the consumer competitive with that from conventional sources. The idea is to foster renewable energy because it generates less air pollution, does not pose long-term waste issues, and is otherwise considered to have far-reaching social benefits.

Green Energy in California accounts for approximately 11% of the total power produced in that state. Five percent of that is from geothermal power alone. It is estimated that 2% comes from biomass and waste burning, 2% from small hydroelectric, 1% from wind power, and less than 1% from solar energy.

The subsidy is boosting the demand for Green Energy, and the state is hoping that with the extra demand, the costs of producing Green Energy will drop so that the subsidy will no longer be needed. Presently, 15,000 residential customers in San Diego County are paying 5% less for this form of energy than they would pay for the energy increment of their power bills from a traditional utility company. The energy increment accounts for approximately one third of the total power bill. This plan is provided by Commonwealth Energy Corporation, a private company that markets electricity to anyone who desires to contract with them.

All of the renewable energy sold by Commonwealth in San Diego County will come from an area of Sonoma County, more than 600 miles north of San Diego, where steam spews from the earth much as it does in Yellowstone National Park. In both areas, tongues of molten lava from deep in the earth reach unusually close to the Earth's surface, causing water tables in some areas to boil and vent as steam. Wells have been drilled toward this heat source, which is being used to heat pressurized water and other fluids. When the pressure is released on the surface, the explosive force of steam or other expanding vapors is being harnessed to spin power generating turbines. All of this goes under the name of geothermal energy.

Now, how does all this Green Energy fit into a book about cogeneration? Well, cogeneration supplies 35 to 40 percentage points of "free" energy when compared to the amount of fuel burned to generate the same equivalent of electricity and hot water conventionally. (See Chapter 1). It would appear that percentage of free energy has the same connotation as Green Energy, i.e. not consuming a nonrenewable resource.

Unfortunately, cogeneration has never gotten the favorable publicity that it deserves. Even our government has ignored this form of energy conservation in its many publications for reasons not explainable. With the deregulation of electricity that allows purchasers to choose from whom they may buy this commodity, it is possible that cogeneration of all sizes will benefit. Heretofore, the electricity, and presumably the heat, from cogeneration had to be used on site or sold to the utility at their rates of purchase. Now, it can be manufactured and sold to anyone who wants to buy it. That means that large users of hot water or steam can support larger cogeneration plants and distribute excess electricity to its neighbors at reduced costs. Small-scale, packaged units in the size ranges of 100 to 300 kW per package are more easily shipped and installed on such sites and may prove overall more economical than megawatt-sized plants, especially when load variations occur.

As one city councilman said when asked about the future of Green Energy, "If you tell the market you want a significant amount of power from green sources, the market will meet the demand. You begin to create competition to produce the kind of energy that ultimately benefits the environment."

One day, cogeneration will be recognized as a form of Green Energy and live up to its proven capabilities.

Chapter 16

Microturbines and Cogeneration

In 1988 two gentlemen, Jim Noe and Robin Mackay, alumni of Garret Corporation, a maker of gas turbines, envisioned a small, high-speed, turbo generator that was uncomplicated, cheap, and not on the market anywhere.

They formed their company and called it NoMac Energy Systems.

From this beginning was developed a 28-kW microturbine generator that has applications in cogeneration, distributed generation, remote power, prime power, and hybrid electric motor vehicles.

The rotor and recuperator section of a microturbine takes up less space than a beer keg and weighs only 165 pounds. Wrapped in a silvery space blanket, it can make electricity from a variety of fuels: natural gas, propane, landfill gas, digester gas, sour gas, kerosene, diesel oil, even gasoline. Running at 96,000 rpm, it is quieter than your vacuum cleaner when you stand beside it. It uses just one moving part, a spinning shaft that serves simultaneously as compressor, turbine, and electric generator rotor. Air bearings support the shaft, eliminating the need for a lubrication system and the attending pumps, reservoirs, and seals. Hence, the unit has capabilities of operating for months on end without maintenance.

The full microturbine package, complete with controls, power electronics, gas compressor (if needed) and air filtration is ensconced in an attractive housing that measures 4 feet long, 33 inches wide and about 6 feet high. The weight of the package is a little over 1000 pounds. Vibration free, this system is ideally suited for close quarters and sensitive surroundings.

The heart of the microturbine is the compressor-turbine unit which is cast as a single piece, curved blades and all, from aluminum and nickel-alloy steel. The design had its origins in the turbocharger field which has enjoyed widespread use in automobiles to force more air into a car's engine to provide added kick when the driver mashes the accelerator.

A combustion chamber mixes fuel and compressed air to provide the combustion gases that enter the turbine wheel at high temperatures. These gases expand through the turbine wheel which provides the power to spin both the compressor and generator which is attached to the same shaft as the turbine wheel. See Figure 16-1.

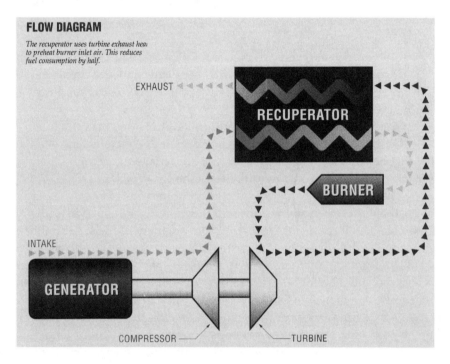

Figure 16-1. 30 kW Capstone MicroTurbine™

The exhaust gases pass through a recuperator, which is a key element in increasing the efficiency of the microturbine. The hot gases exchange their high temperature with the air leaving the compressor causing superheated air to enter the combustion chamber thus increasing the temperature of the gases going to the turbine wheel. This exchange of heat saves fuel thereby increasing the overall thermal efficiency of the microturbine.

Presently, the microturbine is finding acceptability in providing onsite power taking advantage of the deregulation hitting the electrical generation industry. The microturbine supplying 28 kW of electricity is sufficient to provide the air-conditioning power for a large house. Utilities are also enjoying the benefits of placing these pint-sized power plants in areas where peak loads occur to offset having to run less efficient generating equipment to meet demand.

However, the area that might provide the biggest impetus for the microturbine is the hybrid electric vehicle. These include automobiles, buses, and trucks. One of the drawbacks of the electric vehicle is the need to recharge the batteries frequently, therefore reducing its range. By installing a microturbine, which can run on gasoline, diesel fuel, propane or natural gas, the batteries can be charged while the vehicle is being driven down the highway. If this concept should prove successful the quantity of microturbines required to fill this market would mean mass production, thereby reducing the microturbine's overall cost.

Presently, the cost per kilowatt is in the $1200 range. Thus, a 28-kW unit would cost about $33,000. The manufacturer estimates that if production could be boosted to 100,000 units a year, the overall cost per kilowatt would fall to less than $500, or about $14,000 per unit. If that should occur, the microturbine would not only find its other niche in power production, but also in cogeneration applications where the exhaust gases could exchange its heat with water thus producing hot water in quantities that make the microturbine desirable for industrial and commercial users of hot water. But, even at its present costs, the microturbine is proving to be an attractive investment in cogeneration, often showing returns of 25 to 40% depending on the power costs it displaces.

Capstone Turbine Corporation, the successor to NoMac, delineates the applications it is aiming for as:

- Distributed Generation. Expanding incremental generating capacity for electric utilities without constructing new transmission or distribution lines.

- Remote Power. Remote facilities, construction sites, oil fields, marine applications, portable power, developing countries and other locations where utilities are not readily available.

- Prime Power. Supermarkets, stores, office buildings, hospitals, factories, and other buildings with large lighting and air-conditioning loads.

- Cogeneration. Facilities needing electricity and heat where the microturbine exhaust can be used for heating, absorption cooling, dehumidification, baking and drying.

- Hybrid Electric Vehicles. Electric vehicles that need greater range, uphill power and added capability to run heating or air-conditioning features.

- Added power. Avoid Transmission and Distribution costs from a utility when charges for transformers and added lines become too costly.

Capstone has published specifications on their Model C-30 and C-60 MicroTurbine units:

Table 16-1. Specifications

	MODEL C-30	MODEL C-60
Fuel: Natural Gas:	55 psig	75 psig
Full Load Power:	28 kWh	60 kWh
Efficiency:	28%	28%
Heat Rate (LHV)	13,700 Btu/kWh	12,200 Btu/kWh
Emissions:	< 9 ppm NO_x	< 9 ppm NO_x
Noise Level:	65 dba @ 10 meters	65 dba @ 10 meters
Exhaust Gas Temp:	520°F	580°F
Total Exhaust Energy:	3 10,000 Btu/hr	541,000 Btu/hr
Dimensions:	28.1"W×52.9"D×74"H	30"W×77"D×83"H
Weight:	1052 pounds	1671 pounds
Voltage:	400-480 volts Y	400-480 volts Y

Note: Each unit is capable of being both grid connected as well as stand alone, (dual mode). The Model C-30 has a built in gas booster compressor; the Model C-60 requires an external fuel gas booster compressor.

Capstone microturbines, have amassed over 1,000,000 hours of field operation in a variety of applications including cogeneration, oil field service, land fill and digester gas service, hybrid electric vehicles and peak shaving. Over 2500 units have been shipped since December of 1998 through mid 2002 worldwide. Capstone feels that with their patented air bearings and in house recuperator manufacturing, they can quote 40,000 hours of operation before major overhaul.

Other microturbine manufacturers have entered the field in 2001 and 2002, while one, Honeywell Systems has decided to leave the microturbine market place by retrieving their 300 plus units and reimbursing their clients for the trouble. Whether the failure of the GE/Honeywell merger had anything to do with that decision or the fact that the Honeywell machine bad over 25 ppm of NO_x and required much more frequent overhauls that the Capstone unit, is anyone's guess. Honeywell did not announce specific reasons.

Those other manufacturers are:

- Bowman Power Systems is marketing their TG 80 Turbogen microturbine which uses oil bearings, an oil cooled alternator and has a built in heat exchanger for capturing waste heat and producing hot water. Their nominal rating is 80 kWh and they also have a 50 kWh unit of similar construction. Bowman has recently announced a marketing arrangement with Kohler Co. to distribute their products.

- Ingersoll-Rand Power Works has announced their 70 kWh microturbine in 2001 with full marketing via their established air compressor distributors in 2002. The I-R unit is unique in that it has a split shaft arrangement between the compressor wheel and the power turbine wheel. That allows them to integrate a gear reducer in their design to allow for 3600 rpm nominal speed to drive generators and other speeds to drive compressors and pumps. They use oil bearings and incorporate a built in heat exchanger for capturing the waste heat for producing hot water.

- Turbec's T 100 Microturbine is in the market place as of 2002 for nominal 100 kWh production using an external recuperator, but also incorporating a built in heat exchanger for capturing waste heat for producing hot water. Turbec is a joint venture between Volvo and ABB of Sweden.

These established names, Ingersoll-Rand, Volvo, ABB, Bowman/Kohler bring the credibility of the microturbine into focus as a valuable prime mover in the small-scale cogeneration market. While Capstone Turbine is the established leader in this field, it gives them greater credentials in what their efforts have wrought and places the microturbine in a position to reap what experts have been saying over the past 5 years that microturbines, in cogeneration and distributed generation, will capture a significant percentage of the power generation market.

I. Capstone Model C-30 Mictoturbine

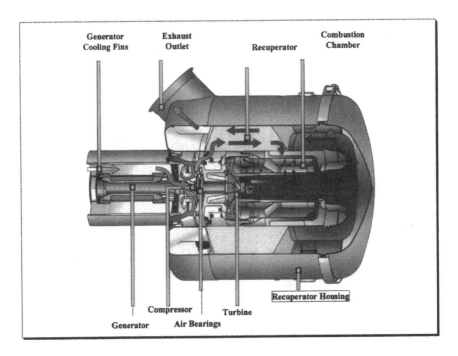

II. Capstone Microturbine cross section

III. Capstone Microturbine one piece rotor assembly showing, left to right the compressor wheel, the power wheel and the permanent magnet generator

IV. The Bowman TG-80 Microturbine

Boiler/Heat Exchanger

V. The Bowman TG-80 showing the built in heat recovery exchanger

Microturbines and Cogeneration

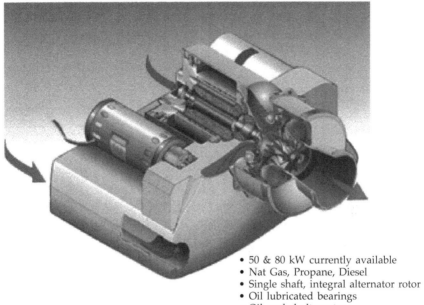

- 50 & 80 kW currently available
- Nat Gas, Propane, Diesel
- Single shaft, integral alternator rotor
- Oil lubricated bearings
- Oil cooled alternator
- Low emissions combustor

VI. The Bowman Microturbine Engine

VII. The Turbec T 100 Microturbine package

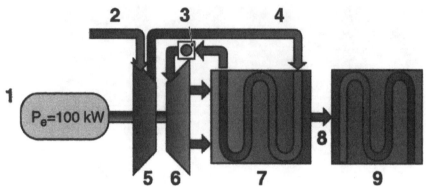

1. Generator
2. Inlet air
3. Combustion chamber
4. Air to Recuperator
5. Compressor
6. Turbine
7. Recuperator
8. Exhaust gases
9. Heat exchanger

Performance at ISO-contitions:
(Net of gas compressor)
Net electrical output: 100 kW
Net electrical efficiency: 30%
Net thermal output: 167 kW
Net total efficiency: 80% (at 50°C WRT)

Noise level: 70 dBA at 1 meter

Emissions, 15% O_2
NO_x: <15 ppmv
CO: <15 ppmv
UHC: <10 ppmv

VIII. The Turbec Thermodynamic Cycle

IX. The Ingersoll-Rand Power Works 70 kWh Microturbine Package

Output   ~ 68 kW
Efficiency ~ 29%? HHV *incl compressor*

Size   3.0 × 5.0 × 7.3 feet high
Weight 3,000 pounds
Cost $??,???

System Cycle Diagram

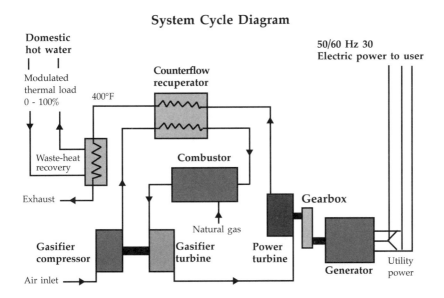

X. The Ingersoll-Rand Power works Split Shaft Schematic

Chapter 17

Absorber-chillers In Cogeneration

The absorber-chiller is extremely suitable for use in cogeneration applications because of its use of heat to produce cooling. Anyone familiar with the gas-fired air-conditioning units offered by Bryant, or the gas-fired refrigerators that are still manufactured and marketed will immediately see how any application of heat, including hot water, may produce a chilling effect.

Furthermore, the absorber-chiller can also produce heat for space-heating purposes as well. The unit is capable, therefore, of supplying both air conditioning and space heating to the building.

Most absorber-chillers use a solution of lithium bromide and water, under a vacuum, as the working fluid. Water is the refrigerant, and lithium bromide is the absorbent. Commercial applications utilize a single-effect chiller-heater and attain cooling capacities of 5 to 10 tons of refrigeration in single units and up to 50 tons in modular assemblies..

The single-effect absorption cycle has a single generator fueled by the exchange between hot water from the cogenerator and a heat medium that is in indirect contact with the dilute solution of lithium bromide and water. In cogeneration applications the conventional gas burner is replaced by hot water at temperatures of between 167 to 212°F. See Figure 17-1 for a typical view of the cooling cycle mode of the single-effect chiller-heater.

COOLING CYCLE

High-temperature Generator

The high-temperature generator heats a dilute lithium bromide solution via the hot water introduced into the heating medium. The boiling process drives the refrigerant vapor and droplets of semi-concentrated solution to the separator.

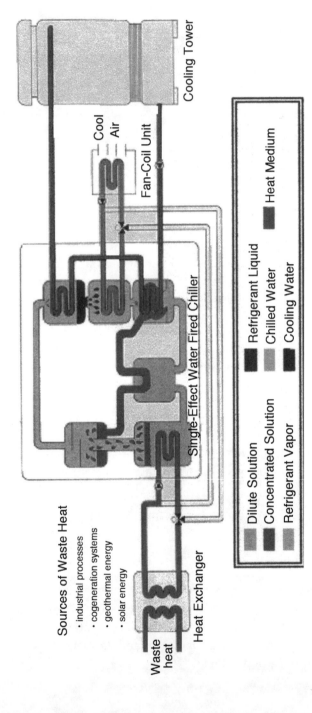

Figure 17-1. Waste Heat Energized Cooling and Heating System (Cooling Operation)

Condenser

In the condenser, refrigerant vapor is condensed on the surface of the cooling coil, and latent heat removed by the cooling water is rejected to a cooling tower, river, or well. Refrigerant liquid accumulates in the condenser and then passes through an orifice into the evaporator.

Evaporator

Pressure in the evaporator is substantially lower that the pressure in the condenser due to the influence of the absorber. As the refrigerant liquid flows into the evaporator, it boils on the surface of the chilled water coil. Heat, equivalent to the latent heat of the refrigerant, is removed from the recirculating water which is chilled to 48.2°F. The refrigerant vapor flows to the absorber.

Absorber

A low pressure in the absorber is maintained by the affinity of the concentrated lithium-bromide solution from the separator with the refrigerant vapor formed in the evaporator. The refrigerant vapor is absorbed by the concentrated lithium bromide solution as it flows across the surface of the absorber coil. Heat of condensation and dilution are removed by the cooling water. The dilute lithium bromide solution is preheated through the heat exchanger before returning to the generator.

Heating Cycle

In the heating-cycle mode, the solution boils in the high-temperature generator, and vapor with concentrated lithium-bromide solution is lifted to the separator in a manner identical to the cooling cycle. Hot refrigerant vapor and droplets of concentrated solution flow through an open changeover valve into the evaporator-absorber. Some refrigerant vapor flows via the low-temperature generator and condenser before reaching the evaporator. Since the pressures in the evaporator and condenser are similar, hot refrigerant vapor condenses on the surface of the chilled hot-water coil. Heat, equivalent to the latent heat of the refrigerant is transferred to the recirculating water, which is heated to 131°F.

In the absorber, liquid refrigerant mixes with concentrated lithium-bromide solution to form a dilute solution and returns to the generator where the cycle is repeated. Figure 17-2 shows the absorber-chiller in the heating cycle.

The features of the Single-Effect chiller heater are:

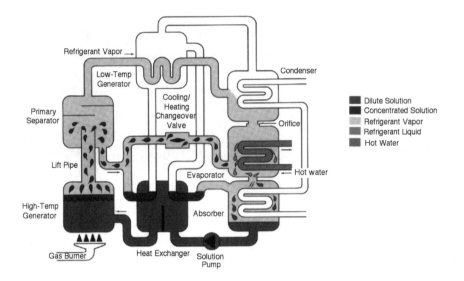

Figure 17-2. Heating Cycle

- Each modular chiller-heater unit serves a dual purpose: cooling and heating

- Safe, odorless, nontoxic lithium bromide and water are the only working fluids and operate under a vacuum at all times.

- Automatic step control on each module increases part load performance.

- Multiple modules are step-controlled in accordance with the cooling or heating demand.

- A two-pipe hydronic system is used to circulate chilled or hot water to a central air-handling unit or multiple fan-coil units.

- Cooling or heating operation can be selected from a remote or built-in switch.

The two most popular manufacturers of small absorber-chillers are Yazaki and Hitachi. The information included in the above description of absorber-chiller operation was supplied by Yazaki as were the figures of the Cooling and Heating Cycles.

Table 17-1 shows the specifications of the Yazaki Absorber-Chiller Models 5-, 7.5- and 10-ton units with the modular assemblies as multiples of those basic models.

In 2002 a direct-fired absorber chiller was introduced in the USA by Broad, USA which is a division of the Broad Corporation out of

Table 17-1. Specifications

		Modular					Modular Assemblies		
Model		5	7.5	10	15	20	30	40	50
Cooling	Capacity (MBH)	59.5	89.3	119.0	178.6	238.1	357.1	476.2	595.2
	Chilled water temp. (°F)*	48.2 Outlet, 57.2 Inlet							
Chilled Water	Rated water flow (gpm)	13.2	19.8	26.4	39.7	52.9	79.3	105.7	132.1
	Evap. press. loss (psi)	7.1	5.1	5.5	5.1			5.5	
Heat medium	Heat input (MBH)	99.2	127.5	170.1	255.1	340.1	510.2	680.3	850.3
	Hot water temp. (°F)	190.4 Inlet, 181.4 Outlet (179.2 Outlet WFC-5)							
	Rated water flow (gmp)	18.4	28.3	37.8	56.6	75.6	113.4	150.9	188.7
	Gen. press. loss (psi)	9.2	2.6	3.9	2.6			3.9	
Cooling water	Heat rejection (MBH)	158.7	216.8	289.1	433.7	578.2	867.3	1156.5	1445.5
	Cooling water temp. (°F)	85 Inlet, 94 Outlet							
	Rated water flow (GPM)	35.2	48.1	64.2	96.3	128.5	192.7	256.6	320.8
	Cond. & abs. press. loss (psi)	7.4	9.5	9.7	9.5			9.7	
Electrical	Power supply	**	208/230V, 60Hz, 1 ph.						
	Consumption (W)	20	30		60		90	120	150
Number of modules			1		2		3	4	5
Weight	Dry (lb)	715	1474	1606	2948	3212	4818	6424	8030

MBH = Btu/hr × 10^3

NOTES: 1. Specifications are based on water in all circuits and fouling factor of 0.0005 ft^{2}hr°F/Btu.
2. A motorized chiller water bypass valve is optional on modular assemblies and requires a 24V, 50/60 Hz (4.8W) power supply.

*Chilled water outlet temperature is factory set at 46.4°F outlet.
**Power supply for WFC-5 is 120V, 60 Hz.

Changsha, China. Broad has produced over 6000 absorber chillers making it the largest absorber chiller manufacturer in the world.

The Broad chiller has a distinct feature in that it is able to take the direct exhaust from a gas turbine (or microturbine) and produce chilled water as low a 41°F. It can also use hot water, steam or natural gas to create chilled water. Southern California Gas Company is presently running a Capstone C-60 microturbine where the exhaust is ducted into a supplementarily fired, 330-ton Broad, USA absorber chiller. The Capstone accounts for about 22 tons while natural gas is burned to produce the remainder. A portion of the chilled water is used to cool the inlet air to the Capstone to a constant 55°F, thereby assuring the Capstone of full 60 kWh production even on the hottest day

Since chilled water is arguably the second most common energy requirement after electricity, the advent of absorber chillers allows cogeneration to be applied in facilities that have little need for hot water or steam. The DOE has a major thrust in the BCHP field (Building Combined Heat & Power) and the use of absorbers in paramount in that thrust. (Read more about that in Chapter 18, Distributed Generation).

The Broad, USA lineup goes from a nominal 50 tons of chilled water to as high as 2646 tons. The units can also produce hot water from the direct fired exhaust up to 195°F. The heating cycle can operate simultaneously during the chiller operation, or hot water can be produced independently using the high stage generator as a very efficient vacuum boiler.

Figure 17-3. Broad, USA Absorber Chiller

Chapter 18

Distributed Generation

The title of this chapter contains words that were not in the cogeneration vocabulary until recently. Much as computer technology has evolved from centralized, mainframe systems to distributed networks of various computing platforms, the traditional model of central-station utility-generating plants is making room for a strategic vision of smaller, distributed resources. This, in anticipation of a deregulated, competitive retail market for energy services.

Today, central-station fossil and nuclear plants supply 87% of the electricity used in the United States. Most of this electricity is generated by coal, oil, and natural-gas fired plants that, on average, are 30 years old. Despite increasing environmental constraints, these aging fossil-fuel-burning plants continue to provide efficient and reliable service. It is uncertain what form of generation will replace these plants as they are retired, although most new capacity in the near term is expected to be high-efficiency, gas-fired, combined-cycle units producing low-cost, commodity-priced electricity. Green Energy, discussed in Chapter 16, will have natural limitations as it strives to harness natural sources of energy production.

Beyond that, beginning around the year 2000, it appears increasingly likely that small, distributed generating units will emerge, initially in niche markets. At the same time, new manufacturing firms will begin to appear that are focused not on large boilers and steam turbines, but on the assembly-line production of microturbines, as discussed in Chapter 15; fuel cells; photovoltaics; and other yet-to-be-developed generating options. This development will signal a change in the power business as revolutionary as the microprocessor was to the computer industry.

No one is predicting the imminent demise of central-station generating plants, which, owing to engineering economies of scale, have long produced some of the lowest-cost electricity available anywhere in the world. But, new technologies for small-scale, distributed generation promise to produce electricity as efficiently as larger plants and, in certain applications, at a cost competitive with centralized generation on a per-kilowatt basis. Referring to Chapter 1 where it was shown that electricity produced by central stations shows a 36% overall efficiency rating, the microturbine, with its accompanying recuperator, is showing almost that much efficiency in 30-kilowatt packages, and by sourcing it at the point of delivery, will lose little in transmission and distribution losses. Combine that with the utilization of cogeneration, producing another work effort, such as water heating, and distributed generation shows economies of fuel conservation as well as economics that will be hard to discredit.

Moreover, many analysts believe that the utility industry's restructuring into regulated distribution companies, independent transmission-system operators, unregulated generating companies, and integrated energy-service providers will create many opportunities for new distributed resources. Couple that with the unbundling of prices for various components of service in contrast to today's simpler, regulated rates, and opportunity has certainly been offered to smart entrepreneurs of distributed generation. (San Diego Gas & Electric, the provider in the author's back yard, has already sold its generating plants to independent operators, changed its name to Sempra Energy, operates as a transmission and distribution company).

Some experts predict that 20% or more of all new generating capacity built in the United States over the next 10 to 12 years could be for distributed generation applications, representing a potential market of several tens of gigawatts.

Conventional technologies for distributed generation range in capacity from tens of kilowatts to tens or hundreds of megawatts. They include reciprocating gas and diesel engines as well as larger gas turbines. Emerging technologies include microturbines of 25- to 75-kW, fuel cells of a few kilowatts to a megawatt or more, and even renewables such as photovoltaics, which may be deployed on individual rooftops at the scale of a few kilowatts as costs continue to fall.

Distributed resources, now being called DR for short in our initialed world, include more than just small generators, however. They also include the backup batteries and other storage technologies that, in many cases, will be coupled with distributed generators to provide ride-through capability during momentary power disturbances and to maintain critical loads during the few seconds it takes to switch from grid power to on-site sources and vice versa. Advanced energy storage systems based on flywheels and ultracapacitors are entering commercial use and could play a significant role. Customer demand and peak-load management technologies that minimize or defer the need for additional electricity from the grid are also gaining momentum in the field of distributed resources.

In theory, distributed resources could lead to far-flung networks of small, interconnected generators and other devices, enabling utility systems to serve growing customer energy needs while minimizing investment in and construction of new central generating and grid capacity. The computer will serve as the motive force to bring multiple small generators together into larger networks that incorporate real-time communication of market prices and centralized automated dispatch.

With the traditional utility approach of "…every kilowatt is ours" no longer applying, utilities themselves will put their vast experience and resources into the picture of applying localized generation to serve the community's needs. The potential of distributed resources in a competitive energy market will depend heavily on the pace, extent, and geographic pattern or regulatory reform across the country.

For example, regions and states where electricity prices are above average (such as California, New York, the mid-Atlantic states and New England) have the greatest potential for DR penetration. Competition and the freedom of retail customers to choose among energy providers may evolve more rapidly in those regions, and retail service companies and other players will enter those DR markets more aggressively.

The most promising users of DR are businesses, factories, and various other sites needing steam or hot water from cogeneration. Distributed generation technologies like fuel cells and small gas turbines are well suited for such sites. These and other DR's can also provide economical peak shaving, high power quality and standby and uninterruptable power. Present restrictions from generating too much power and not enough thermal energy will be mitigated by either redefining PURPA or eliminating the need for this regulatory body. The economics of the DG will be the guiding light as to whether a cogenerating system will continue to operate when the need for thermal energy is at an ebb. Sell-back agreements will be passé since the seller will also be the buyer.

STRUCTURAL CHANGES DRIVE INTEREST IN DR

The Electric Power Research Institute, EPRI, has identified three structural shifts in the electric power industry that are driving interest in distributed resources. First, the shift away from command-and-control market regulation will result in the deregulation of about 70% of the utility sector's total economic value. The generation, trading, and retailing segments of the industry will become much less regulated, while Transmission and Distribution (T&D) functions will be more regulated.

Second, an enormous transfer of generation assets from the regulated rate base to unregulated enterprises is under way. At least 15 utilities have already announced plans to divest themselves of more than 25 Gigawatts (GW) of generating capacity. By 2001, fossil fuel plants totaling 100 GW could change hands; by 2006, some 400-450 GW of capacity could be spun off from the rate base of investor-owned utilities, most of it sold through auctions.

These shifts will, in turn, lead to a third major shift: the opening of a $500 billion, competitive wholesale and retail market that will bundle electricity and gas with other household services (e.g. water and cable communications) and such ancillary services as facility management and performance contracting.

As a result of these major changes, a new value chain will form and new business models will evolve to seize opportunities created by technology innovation. Ideas will be gleaned from other industries that are already largely deregulated, such as telecommunications, natural gas and financial services. This value chain will be much larger than the power industry alone and will aim to serve a $1-trillion customer base for all the services, energy, telecommunications, and infrastructure, needed by people and organizations to operate in buildings. There are plenty of signs that customers want to buy bundled energy and infrastructure services—witness the popularity of total customer solution marketing in many other service industries.

This new value chain is likely to have five principal segments:

1. *Generating Companies* will focus on producing power under market conditions.

2. *Transmission Companies* or network operators will run the high-voltage systems for long distance power transfers.

3. *Trading Companies* will buy power from generating companies and serve as intermediaries to sell power to downstream customers.

4. *Local Distribution Companies* for electricity, gas, telecommunications, cable service and the like will be the platforms for delivering electrons, Btu's and other services.

5. *Retail Merchant Companies*, meanwhile, will sell the services to customers, buying wholesale energy from traders and getting it to customers through the distribution companies.

NEW OPPORTUNITIES IN THE VALUE CHAIN

The emergence of such a value chain will coincide with three other major trends. First, there will be both consolidation and fragmentation in the power industry. The ownership of large, centralized generating plants will be consolidated in order to cut production costs. At the same time, new opportunities will appear for localized and distributed generation in medium unit sizes to complement this bulk-supply system and to serve as a hedging asset to balance supply and demand portfolios.

Second, there will be an outpouring of new service offerings to meet the needs of new wholesale and retail customers. Wholesale market offerings will tend to be structured; in the retail market, some customers will want customized solutions, while most may prefer standardized, one-rate service plans.

Third, new technologies, both for generation and for trading and transaction processing, will be used to help serve these new customers. Technology will profoundly affect the way electricity is generated and brought to the competitive market.

On the generating side, there will be plant upgrades and the re-engineering of operation and maintenance activities; enhanced real-time diagnosis and control; advanced gas turbines and more flexible operating cycles; and improvements in distributed generation, including more types of prime movers, better controls, lower costs, and increased reliability.

On the trading and transaction side, faster trading systems and bargain-hunting expert systems will play a role, as will advanced customer knowledge systems and the growing sophistication of service offerings.

The new value chain envisioned will involve three forms of power:

1. Commodity-priced bulk power for wholesale transactions at the power pool or interpool level will be generated by large baseload and intermediate sized plants.

2. The second type of power, for local purchase by distribution and retail merchant companies, is called logistical power and will be generated in smaller amounts and for shorter periods to help balance wholesale supply and retail demand. It will be produced by specialized generation companies operating within regional electricity networks. In each region, half a dozen such specialized logistical producers may emerge, each supplying 40 to 50 energy companies. These producers will offer back-up reliability services, peak management contracts, and price risk containment insurance.

3. The third type of power, retail power, will be sold at the end-user level as part of a broader offering (bundled with other commodities and combined with facility services). Some of that power will be generated at customer sites. Eventually, the regional logistical companies mentioned above could offer transmission & distribution bypass and arbitrage services, using networked and other interconnected forms of distributed generation to compete with on-site distributed generation.

Industry restructuring and the emergence of a new value-added chain in the production and delivery of electricity is likely to lead to an explosion in electricity-product innovation. Distributed generation will offer the means for providing more profitable, value-added custom-service offerings in the intermediate wholesale and retail markets.

FOCUS ON CUSTOMERS

One upshot of industry restructuring along a new value-added chain is the emergence of an entirely new set of potential DR customers.

Merchant retail companies may develop local networks of distributed generation to support their service business.

Distribution companies could become customers for distributed generators in the 1- to 2-MW range for operations support.

Energy-demand aggregators may want to own interests in similar size units to help firm up loads for better prices.

Logistical companies may own some cogeneration or small, dispersed power units to balance asset portfolios and provide flexibility to power-contract transactions.

Given the anticipated involvement of these types of players, end users are expected to account for no more than 15-25% of the total amount of distributed generation that is installed. But even that percentage range could translate into very large numbers of installations in various types of buildings. EPRI has analyzed a significant number of key market segments in detail and has identified a potential of 1.6 million existing establishments, representing an aggregate load of 288 GW and annual revenues of some $66 billion, where distributed resources conceivable could be economical.

DR penetration of these markets is likely to occur gradually. But, by the time retail-electricity markets are fully de-regulated (around 2002-2003), the equivalent of 25-30 GW of load may be accessible to distributed generating units in the 200-kW-to-1-MW range. During the transition to full retail competition, some dozen or more utilities may invest in on-site generating facilities in applications of 1-2 MW as part of a strategy to retain larger customers, creating a near-term market for distributed generation of as much as 2 GW.

Regulatory treatment of current utilities' stranded costs will be at issue along with pollution emissions, therefore the implementation of many distributed generators will be scrutinized to evaluate the impact on those two factors.

There are other issues facing distributed generation as well. The need for less expensive, standardized switchgear is one such issue. Another is the need for broadly accepted interconnection standards that ensure personnel safety and the protection of customer owned equipment from distribution system anomalies. Power-conversion technology must be addressed to allow for the direct-current output of some distributed generators such as fuel cells and various turbogenerators which need their power converted to grid-quality alternating current.

A FLEXIBLE ENERGY OPTION

Although opinions vary widely about how rapidly and how extensively distributed resources will be developed and deployed in the United States during this period of industry transition, there is little disagreement about their virtually infinite potential in developing countries that have little or no existing power delivery infrastructure.

Moreover, in Europe, rapidly growing customer markets for cogeneration are providing opportunities for distributed generation. The deployment of distributed resources abroad is expected to help reduce costs, which in turn will make the technologies more competitive in the U.S. markets.

Worldwide, the convergence of market competition, customer choice and the undeniable benefits of electrification are making future prospects for distributed generation brighter than ever.

Traditional utilities are seeing a significant cultural change, and those that embrace those changes will transform their businesses from century-old electric utilities into competitive energy companies and service providers. Those that look at distributed generation as the competitive threat that it is and fail to modify their normal *modus-oprandi* will be bypassed as power will be available from remote sources as well as more local sources. Utilities that once looked at cogeneration as a theft of their domain power will be further surprised at the implementation of cogeneration at all levels of capacity as long as there is need for the thermal by-product. Distributed generation will finally be the boon for cogenerator manufacturers who have been thwarted by archaic laws and utility indifference as tens of thousands of users that have cogeneration requirements will be allowed to implement them freely.

Only time will tell how successful DG products and services will be in penetrating the energy market. There are many challenges to be faced, but distributed resources will give energy companies and energy users a flexible energy solution that can complement electricity straight from the grid.

Chapter 19

United States Combined Heat & Power Association

A major effort has been started to double the amount of power derived from cogeneration by the year 2010, The driving force behind this effort is the U.S. Combined Heat & Power Association, Therefore, on any book about cogeneration it is incumbent upon the author to include the efforts of this organization in a separate chapter like this.

This organization has the full assistance and attention of the U.S. Department of Energy (DOE) and the Environmental Protection Agency (EPA).

As of 1998 the production of electricity from combined heat & power plants, cogeneration, was 46 gigawatts (GW). A gigawatt is equal to one billion watts or 1000 megawatts (MW) or one million kilowatts (kW). The USCHPA's goal for the year 2010 is to increase that to 92 gigawatts. Presently, cogeneration accounts for 7% of the nation's total generating capacity up from 2% in 1990. Cogeneration, combined heat and power, represents 40% of the non-utility generating capacity. This latter figure is a little misleading since it represents some of the private power producers purchase of utility plants in the advent of deregulation. If this goal is attained, the percentage would not necessarily increase to 14% of the nation's total generating capacity since that overall total will have risen due to new, non-cogenerating, power plants that will be built to meet the needs of this power consuming nation.

The suggested areas in which these additional gigawatts can be added are broken down by the USCHPA as follows:

- 27 GW of additional industrial CBP capacity by replicating "best practices," supporting the use of output-based emissions standards in more states and by EPA, and in participating in cost shared R&D projects with Federal and state government agencies in the areas of advanced industrial power generation, black liquor and biomass gasification, advanced materials and combustion processes, and advanced power electronics, communications and controls.

- 8 GW of additional buildings cooling, heating and power capacity by implementing the BCHP Roadmap; conducting a coordinated outreach campaign to educate architects, building designers, and local building and other code officials about BCHP; providing "SWAT" team technical assistance to those interested in installing BCHP systems, and participating in cost-shared R&D projects with Federal and state government agencies in the areas of packaged system integration, power electronics, communications and controls, fuel cells, microturbines, reciprocating engines, and thermally activated cooling and humidity control equipment.

- 8 GW of additional district energy capacity by expanding education and outreach efforts to municipal and community governments, college campuses, and military bases; providing "how-to" guidebooks to those interested in installed district energy systems; and advocating more demonstration projects of innovative applications in power parks, communities, "brownfield redevelopment" and public housing projects.

- 5 GW of additional CHP capacity in federal facilities by working with the Federal Energy Management Program (FEND) and federal sites to identify new sources of funding for the installation and operation of CHP systems; conducting assessments of CUP opportunities in federal facilities nationwide; working with FEMP to provide technical assistance to facility managers interested in installing CHP systems; and conducting case studies to demonstrate all forms of CHP in Federal facilities across a wide range of building types, agencies, and regions of the country.

The U.S. Combined Heat & Power Association workshops have identified the major problems we are facing today:

- Energy Prices
- Power Outages
- Power Quality

- Dirty Air
- Global Climate Change

Cogeneration can allow us to make progress in solving all these problems.

The USCHPA workshops identified these critical issues, but one point stands out: Because of problems in energy markets today, unless action is taken soon, the progress America has made over the last decade in the economy and environment could stall or even reverse. Demand is outstripping supply. Combined Heat & Power is one of the most cost effective sources of clean energy generation. America need CUP more than ever before.

The present power grid consists of approximately 15,000 power plants producing over 3 trillion kilowatt hours of electricity with a reliability factor of 99.7%. That is almost three "9's" in the new reliability nomenclature of the 21st century. It means that the average electric buying customer will suffer 8 hours of power outage per year. Hospitals require at least four "9's" of reliability; while e-commerce is demanding a minimum of six "9's" which is an outage of 30 seconds per year. The current system cannot provide that reliability. On-site generation is needed to achieve the power reliability that the modem information age demands.

Furthermore, power quality issues are almost as important. Power surges and sags cannot be tolerated in our computer driven world. Poor power quality is capable of damaging sensitive electronic equipment. On-site power generators can solve the problem of power quality.

The aging infrastructure of our transmission and distribution lines creates problems of reliability even if the power is being generated but cannot get to the users due to line breakage or overloading.

The workshops identified and expanded upon the benefits of cogeneration:

- CUP is a win-win-win situation for energy users, equipment and energy suppliers, and society in general
- The potential is enormous for industry, commercial buildings, federal buildings and district energy organizations.
- There is a need to resolve certain regulatory and institutional barriers such as grid interconnection, environmental siting and permits, utility policies, local building codes and tax treatment.

- Barriers to CHP involve regulatory and/or policy solutions, therefore there is a need for industry/government partnership to resolve these issues.

- Identification of rules and regulations between participants is needed.

Actions that need to be taken include insuring that awareness of the problems is adequately disseminated, removing regulatory and institutional barriers and enhancing the technology and market development of the products that encompass CHP.

Starts have been made in all these areas with great success in various parts of the country. California, New York and Texas are among states that have pre-certified on-site generation equipment for interconnection standards. That means that the wheel doesn't have to be re-invented every time application is made to a utility for interconnection. Proof that IEEE standards are adhered to further makes both interconnection and building permits easier to attain. Fair utility practices and policies are needed to address standby charges, exit fees and competitive transition charges. California passed legislation to suspend stand by charges which are charges utilities impose upon on-site generators at some dollar per installed kilowatt. Those charges range from $1 to as high as $7 per kilowatt, making a 300-kilowatt system pay up to $2100 per month just to be interconnected to the utility. Exit fees are charges imposed on a per kilowatt hour basis for no longer buying electricity from the local utility. Competitive transition charges are fees that allow the generating utility to recoup any loss of invested capital due to rate freezes before that utility becomes fully de-regulated.

But utilities are not the only barriers. Emission standards must be reasonable with the knowledge that on-site cogeneration uses less fuel overall thereby reducing the total emission of pollutants. Localities need to streamline siting and permitting issues and practices to prevent every on-site generator to be meticulously introspected under such guises as "design review" and "planning commissions" which are time and money consuming practices. Safety issues are paramount and most communities already have city and local codes that address those areas. Permits should insure that the installations meet local codes, but the process of initiating construction should not be held up because a committee of city planners need to study whether existing codes are sufficient to impose upon a new project.

Equitable tax practices need to be looked into to allow faster depreciation of equipment for the benefit of the developer or owner of on-site cogeneration. The California PUC has instituted a rebate program that for cogeneration projects under 1.5 megawatts that meet minimum efficiency standards (similar to FERC standards) a $1/watt or 300/6 of the overall construction cost, whichever is lesser, will be given to the owner after project completion. This program is slated to run through 2004 and is administered by four regional energy agencies which are coupled to the three major public utilities and the So Cal Gas Company.

Table 19-1 shows the National CHP Roadmap to reaching these goals.

Table 19-2 shows the Action Plan for Eliminating Regulatory and Institutional Barriers.

Uniform grid interconnection standards are sorely needed to solve the problems many project developers have experienced interconnecting with the utility grid. And, these are not just limited to CUP installations. Many on-site and distributed energy generation projects—e.g., rooftop photovoltaic installations, wind turbine projects, industrial self generation systems, and backup power supplies—encounter similar interconnection difficulties. Smaller projects that cannot bear the exorbitant costs of utility interconnection fees find that these costs are "deal breakers." Utility interconnection requirements often go beyond the minimum standards needed to insure safe and reliable grid operations. Requirements vary across service territories and state and have been known to vary on a project-by-project basis. The market for CHP will not develop on a large scale until there is a national solution to the interconnection issue.

The IEEE Interconnection Standard—P1547 should go a long way to alleviating any fears by the utility companies that dangerous conditions will prevail if interconnection without "red tape" is the mind set. Once this standard is adhered to by the suppliers of on-site generators, and documented to the satisfaction of the utility company, interconnection should be made easier and less costly to attain.

EPRI and the DOE have published reports documenting interconnection problems. Some of the utilities policies and practices whose effect is to place severe limits on the viability and cost effectiveness of CHP installations are:

- Direct prohibition by the local utility from operating and interconnecting an on-site CHP system in parallel with the grid.

142   Small-scale Cogeneration Handbook

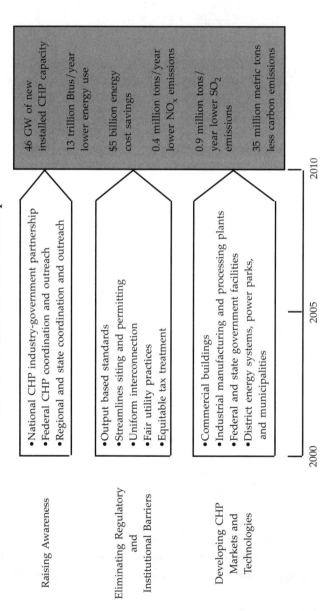

Table 19-1. National CHP Roadmap

Table 19-2. Eliminating Regulatory and Institutional Barriers

The CHP Vision outlines a number of regulatory and institutional barriers that interfere with the expanded use of CHP systems in the U.S. The presence of these barriers was confirmed at each and every one of the regional and state CHP workshops. The elimination of these barriers is the centerpiece of the National CHP Roadmap.

The most pressing regulatory and institutional barriers facing CHP include the following:

- Irregular interconnection requirements
- Unjustified and costly standby and backup power charges
- Prohibitive stranded cost-recovery charges and exit fees
- Air regulations that do not recognize the environmental superiority of CHP
- Irregular environmental permitting procedures
- Time consuming and confusing site permitting (e.g., zoning, building, fire, and safety codes)
- Inconsistent tax treatment and depreciation policies

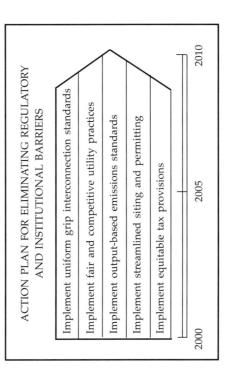

ACTION PLAN FOR ELIMINATING REGULATORY AND INSTITUTIONAL BARRIERS

- Implement uniform grip interconnection standards
- Implement fair and competitive utility practices
- Implement output-based emissions standards
- Implement streamlined siting and permitting
- Implement equitable tax provisions

2000 — 2005 — 2010

- Utility tariff provisions that are seen to discourage CHP, such as demand charges and backup rates, buy-back rates, exit fees, "uplift" charges" and competitive transition or stranded cost recovery charges.

- Transmission access procedures, rules and costs.

- Selective discounting of utility services to large customers to prevent the use of on-site generation.

Coordinated efforts need to be undertaken on a national basis to address and eliminate utility policies and practices that unnecessarily discourage distributed energy and CHP projects. Organizations such as the Distributed Power Coalition of American and the California Alliance on Distributed Energy Resources (CADER) arc working with the National CHP Roadmap to implement sound interconnection standards. Other organizations such as the National Association of Regulatory Utility Commissioners and the National Conference of State Legislators are critical partners in accomplishing the following actions:

- Develop and promulgate **standard commercial practices** and business terms for utilities in their dealings with distributed energy and CHP developers.

- Develop and disseminate **"model" utility regulatory principles**, tariffs, and legislative provisions for DG and CHP projects.

- Develop **analysis tools, data, and case studies** for assessing the value and impacts of DG and CBP systems on local electricity and natural gas distribution systems.

- **Establish dispute resolution process and capabilities** for expediting DG and CHP project proposals.

CHP must, by its very nature, be localized. It falls under the overall heading of Distributed Generation and this chapter could have easily been incorporated into the previous chapter on DG. The amount of heat generated by the central station utilities cannot be readily used due not only to the size of the plant but the distance that most plants are located from where the heat can be utilized—the district energy concept. Ironically, Thomas Edison's use of DC current production from his Pearl Street central plant made it necessary that the plant was in the midst of civilization since DC current could not be transported very far. The subsequent waste heat from his prime mover was easily distributed to the surrounding homes and industries. When Tesla and Westinghouse devised alternating current, AC, central stations could now transport electricity great distances and plants got bigger and moved away from the heat using public. Presently, the United States district energy systems account for about 3% of the total electrical production. That compares with 70% in Russia, 50% in Denmark and 44% in Sweden. Those countries are cold climate countries where the necessity for inexpensive heat was mandatory and district energy systems made the most sense. The United States cheap energy blessings thwart the urgency for district energy. Now that energy is becoming more expensive in major sectors of the country and the "national grid" is causing other sectors to see rising electricity prices, CUP has the attention of energy suppliers and users alike.

The USCHPA has also fostered a new initiative called BCHP or Building Combined Heat & Power. Presently commercial, institutional and multi-family buildings consume one-sixth of our total energy production. There is over 60 billion square feet of building space in the United States. Presently, four states—New York, California, Texas and Pennsylvania account for 50% of the combined heat and power utilization in buildings. Virtually all of this CUP in commercial buildings is in the form of absorber chiller air conditioning since very little hot water is used in the typical commercial buildings. Institutional buildings can use sufficient hot water to make CUP economical in providing both hot water and chilled water. The same holds true for multi-family buildings, but often separate metering of electric usage in those buildings causes difficulty in interconnecting any electrical production except for the common areas.

The USCHPA has joined several other industry trade organizations in the implementation of this **National CRP Roadmap**. In addition, other industry organizations, private businesses, national laboratories, and non-governmental organizations have been active participants. This coalition's key organizations included the following:

- American Gas Cooling Center (AGCC)

- Buildings Cooling, Heating and Power (BCHP)
- Distributed Power Coalition of America (DPCA)
- International District Energy Association (IDEA)
- Gas Technology Institute (GTI)
- American Gas Association (AGA)
- Council of Industrial Boiler Owners (CIBO)
- National Association of Energy Services Companies (NAESCO)
- International Federation of Industrial Energy Consumers—North America (IFIEC)
- American Forest and Paper Assoc. (AF&PA)
- Solar Turbines
- Trigen
- Onsite-Sycom
- Dow
- Duke Solutions
- Energetics
- Exergy Partners
- Keyspan
- Mississippi Valley Gas Company
- NiSource
- Northwind Boston
- Southern California Gas Company
- Tecogen
- Weyerhauser

Participating non-governmental organizations and national laboratories include:

- American Council for an Energy Efficient Economy (ACEEE)
- Alliance to Save Energy (ASE)
- Northeast-Midwest Institute (NEMW)
- Oak Ridge National Laboratory (ORNL)
- National Renewable Energy Laboratory (NREL)
- Pacific Northwest National Laboratory (PNNL)

The U.S. Combined Heat & Power Association via its National CHP Roadmap has made and will continue to make deep inroads to the vast implementation of CHP projects across our country. Economics will still be the driving force and those areas that are blessed with very cheap electricity may have to stretch their boundaries, but any technology that increases the fuel utilization efficiency from 35% to 85% must be looked at regardless of present cost of electricity. As a rule of thumb, on-site generated electricity in today's natural gas price economy will work if the cost of electricity exceeds 5 cents per kilowatt hour.

The USCHPA believes that if the goals of doubling the amount of electricity produced via cogeneration are reached—96 gigawatts by 2010—then the following will result:

- $5 billion in energy cost savings
- 1.3 trillion Btu's/year in energy consumption savings
- 0.4 million tons reduction in NO_x
- 0.9 million tons reduction in SO_2
- 35 million tons of volatile carbon reduction

It behooves all applications of combined heat & power be explored thoroughly to attain these astounding results.

Chapter 19

Cogeneration in Europe

In every major European country, there is a nonprofit organization promoting the usage of cogeneration and acting as a platform for the various interests involved. These national institutions are members of Cogen Europe, a nonprofit organization based in Brussels, Belgium, whose main focus is to promote cogeneration on a multinational level.

The European engineers take a different approach to designing cogeneration plants. Instead of building large gas turbines, or combined-cycle plants whose main target is to produce electricity, and then trying to utilize as much heat as possible, European engineers target the replacement of the base heat supply of certain, small entities. By focusing on the annual heat-demand graph, the basic layout for maximum utilization is determined. If a plant can use all or a majority of the electricity, the "by-product" produced in this combined process, the perfect requirements are a given. While no European country has the equivalent of FERC or PURPA, this approach is very similar to the requirement in the United States to insure a Qualified (Cogeneration) Facility meets minimum FERC efficiency in order for it to be considered a PURPA plant and derive the benefits of utility interconnect and standby power as well as qualified fuel rates.

Producing heat and electricity at the point of need—that means creating an end of the pipe solution—is a very effective way to ensure engineering keeps the optimization of the relation between demand and production as its main target.

One major philosophy that some governments in Western European nations utilized was to provide cogeneration investors with subsidies and tax breaks. These subsidies caused the number of cogeneration installations to increase which had the effect of having cogeneration technology become more advanced and less expensive. All these factors started a cogeneration boom in the late 1980's and early 1990's in certain Western European nations.

Two distinct factors prevailed causing this increase in cogeneration investment:

1. Social and Political Factors.
   - Public pressure against large power plants.
   - Emissions regulations.
   - Energy saving strategies.
   - Deregulation and liberalization.

2. Economic Factors.
   - General high energy prices.
   - Attractive gaps between electricity prices and primary fuels.
   - Decreasing investment and operations costs.

Today, cogeneration is one of the prime technologies available to achieve two valuable goals:

1. Efficient usage of limited resources.
2. Air pollution reduction.

With the United States importing over 50% of its oil, one would think the U.S. would have taken similar measures that the Western European nations have implemented. But, one hardly hears of cogeneration from the Department of Energy. And, there are certainly no tax breaks or subsidies for cogeneration projects coming from the Federal Government. The last major tax break for any capital project was the investment-tax credit that applied to any capital investment. This was instituted in the early 1960's, phased out after about 10 years, and has never been heard from since.

SOME IMPRESSIVE NUMBERS

Denmark, Finland, and The Netherlands are the European leaders in the ratio of cogeneration to total national power production. In these countries, between 30% to 40% of their total power production is provided by cogeneration. In the European Union an average of 10% of total power production is done through cogeneration systems.

In the last 10 years, Germany and The Netherlands installed approximately 10,000 cogeneration units based on gas engines and turbines. In Germany, alone, the number of engine- and turbine-based cogeneration systems grew from 500 plants to over 3300 plants from 1985 to 1995. In these 10 years, almost six times as many installations were made than existed before 1985.

FUTURE OUTLOOK

The only consistent factor in the European industrial power market is rapid change, which can be viewed as both positively and negatively for equipment manufacturers. The key areas of change which can have dual effects on the markets are:

- Industrial production cost reduction
- European Union liberalization of power markets
- Gas prices and infrastructures
- The rise of contract energy management
- Technical advances
- Environmental policy

Despite all the possible negative influences most experts see a bright future for the Independent Power Producer (IPP) market.
The existing prognoses state that the European Union will triple their power production from IPPs by the years 2005 to 2010. This means that within 6 to 11 years, an average of 30% of power production in Europe will be generated by cogeneration. Presently, of all the power plants built today, more than 30% are IPPs.

The main reasons for this growth are both political and technological. Based on the recommendations made by Cogen Europe, the European Community is promoting cogeneration to account for at least 20% of total European power generation.

The EU Environment Ministers agreed in 1997, that a collective 15% reduction of the greenhouse gas emission should be achieved by 2010. This was based on the emission levels in existence in 1990. In order to achieve this reduction in emissions, the ministers set up a recommendation list for a wide range of policies and measures that can be used at a community-wide level to help achieve these targets. Cogeneration is ranked highly on this list of recommended policies and measures.

Improving the specific investment costs and operating costs dramatically through the development of better technologies guarantees a much better ROI which will attract more and more investors. In 1990, the investment cost of a gas-engine-driven cogeneration plant in the 5-MW range was approximately $1,000/kW. Today, a similar plant would cost approximately $700/kW, installed. There is no reason not to expect similar reductions in small-scale-cogeneration technology, especially with the advent of the microturbine as its effect is felt in cogeneration.

In addition to having lower investment costs, today's systems have better efficiency and lower maintenance costs than those systems installed just five to seven years ago. Engine and turbine technologies have made tremendous developments in the recent years showing that these technologies are both ecologically and economically the prime choice for power generation if the right conditions exist.

A risk analysis considering all the factors that the deregulation of the electricity markets will affect, shows that the clear winners of such a scenario are:

- IPPs
- Renewable-power producer
- Energy brokers

ESTIMATE MARKET FIGURES

Based on the numbers of 1995 and 1996 as well as on the forecast of installations for the next 15 years, the cogeneration market in Europe appears to have impressive potential. Major participants will be the power-equipment manufacturers. It is very likely that within the next few years a consolidation wave among power equipment manufacturers occur with many mergers and acquisitions. Table 19-1 shows estimates of the sales volume of the major providers in this market.

Table 19-1

	Approx. Sales per Year
Power-equipment Manufacturers - Engines	$1.8 - 2.5 billion
Power-equipment Manufacturers - Turbines	$8.0 - 12.5 billion
Accessory Equipment	$3.0 - 4.5 billion
Maintenance and parts	$1.0 - 1.5 billion
Engineering and Consulting	$0.7 - 1.4 billion

Not included in these market numbers, but the group of companies which most likely will see a tremendous business forced by the cogeneration market development, are the gas companies. The forecast for Germany only shows that the existing gas consumption for power production will double by year 2010 to over 32 billion cubic feet from 16 billion cubic feet in 1995. The growth from 1985 to 1995 was only 60%.

What can U.S. companies do to take part in this boom that is presently occurring in Europe? There are 10 rules that may form the basis for their marketing.

1. Choose entry strategy carefully
2. Perfect time for acquisitions and joint ventures
3. Partners provide easier access to European markets
4. UK and The Netherlands as prime markets for cogeneration
5. Seek professional assistance
6. Utilize free sources of information
7. Avoid tight budgets
8. Have dedicated area managers in charge of specific regions
9. Control flow of information
10. Do not underestimate the different ways of doing business abroad

While it took the energy crises of the 1970's for societies to become aware of limited energy resources, those nations and companies that understood that the word crisis is derived from the Greek words for danger and opportunity are certainly reaping the socioeconomic benefits of getting their energy houses in order.

APPENDICES

Appendix I

Typical Shared Savings Agreement

Cogeneration Energy Purchase Agreement

Section 1. Parties and Effective Date

1.1 **Parties to Agreement:**
This Cogeneration Energy Purchase Agreement ("Agreement") is entered into between (i) **[Customer Name]** ("Host") and (ii) **Third Party Financier** (TPF). This Agreement shall be binding on the permitted successors, assigns, and transferees of the Host and TPF.

1.2 **Effective Date:**
This Agreement shall become effective on the date ("Effective Date") on which (i) it has been approved by the authorized representatives of both the Host and TPF. From and after the Effective Date, TPF shall exercise due diligence to obtain, at its sole cost and expense, the approval of **[Local Electric Utility]** and all other governmental and non-govern mental agencies as are necessary for the installation and operation of the System (hereinafter defined).

Section 2. Recitals

2.1 **Recital 1:**
The Host is the owner and operator of the **[Customer's Facility]** ("Facility") located at **[Customer Address]**. Host desires to purchase from **TPF** a portion of its electrical and thermal energy to provide electrical and heating service to the Facility.

2.2 **Recital 2:**
TPF desires to design, construct, install, own, maintain, and operate at the Facility a cogeneration system ("System") for the production of electricity and thermal energy. A description of the System is attached as Exhibit A. Host and TPF have agreed upon the installation plans and specifications for the System.

Section 3.   Terms

3.1 **TPF Representations and Warranties:**
TPF represents and warrants to the Host that it has conducted a detailed audit of the Facility to determine the amount of electrical and thermal energy that may be required to provide electrical and heating service to the Facility ("the Facility Audit"). TPF represents and warrants that the System, as designed, installed, and operated, will be capable of producing a portion of the Host's electric energy and thermal energy in accordance with the functional specifications of the System and the manufacturer's faceplate for outputs of the System. The Host acknowledges and agrees that the System design is based on the Facility's physical configuration, estimated energy requirements, and other facts, estimates and assumptions identified in the Facility Audit dated _____, and the Host acknowledges and agrees that the System as designed is suitable for the Host's purposes.

Except as provided in Section 3.10, TPF shall not be responsible for any work done by others unless authorized in advance by TPF. TPF shall not be responsible for any loss, damage, cost, or expense arising out of or resulting from improper environmental controls, improper operation or maintenance, fire, flood, accident or other similar causes. If TPF determines, in its sole discretion, that a problem is not covered by this warranty, Host shall pay TPF for diagnosing and correcting the problem at TPF, standard rates.

THE WARRANTIES AND REMEDIES SET FORTH ABOVE ARE EXCLUSIVE, AND NO OTHER WARRANTY OR REMEDY OF ANY KIND, WHETHER STATUTORY, WRITTEN, ORAL, EXPRESS, OR IMPLIED, INCLUDING WITHOUT LIMITATION WARRANTIES OF MERCHANTABILITY AND FITNESS FOR A PARTICULAR PURPOSE, OR WARRANTIES ARISING FROM COURSE OF DEALING OR USAGE OF TRADE

SHALL APPLY. The remedies set forth in this Agreement shall be the Host's sole and exclusive remedies for any claim or liability of any kind arising out of or in connection with this Agreement, whether arising in contract, tort (including negligence), strict liability or otherwise.

3.2 **Design, Construction and Installation of System:**
TPF shall, at its sole cost, expense, and liability, design, construct, install, own and maintain the System at the Facility. TPF shall provide and install a sound attenuation cabinet and foundation capable of adequately supporting the installation and operation of the System.

TPF, shall install the System and shall provide supplemental electrical and thermal service to the Facility within 180 days of the Effective Date.

3.3 **Host's Representations and Warranties:**
TPF shall be entitled to rely on the accuracy of any and all information provided by the Host, as it pertains to the Facility's physical configuration, estimated energy requirements, and other facts, estimates and assumptions, as identified in the Facility Audit dated _____, which is hereby warranted by the Host to be accurate and correct. For all equipment to be installed at the Facility, the Host warrants that it either owns the Facility, or has the authority to permit TPF to install the specified equipment on the Facility. In the event of any unforeseen difficulties in installing or operating the System due to conditions at the Facility or due to the inaccuracy of any information relied upon by TPF, the price, schedule and other terms and conditions of this Agreement shall be equitably adjusted to compensate for any additional work required or performed by TPF.

The Host shall be responsible to provide suitable space that meets all environmental specifications for any equipment to be installed at the Facility. The Host shall also be responsible to provide safe access to the Facility as necessary for TPF to install, operate and maintain the System. The Host shall compensate TPF for any delays or additional work that becomes necessary because of inadequate access to the Facility or work area at TPF then standard rates. The Host shall ensure that all Occupational Safety and Health Act requirements are adhered to for the areas where the System is to be installed and operated, and where any TPF equipment used in operating and maintaining the System is to be stored. In the event of damage to TPF equipment that is caused by Host or Host's agents, Host agrees to pay all repair or replacement costs associated with the damage. TPF shall have the right to suspend work in the event that there is inadequate access to work area, or in the event that the safety of any person or property might be jeopardized by continuing with the work.

3.4 **Ownership of System:**
The System and all plans and specifications related thereto shall at all times be owned by TPF.

3.5 **Term of Agreement:**
The Initial Term ("Initial Term") of this Agreement shall commence on the Effective Date and shall terminate eighty-four (84) months after the date on which TPF first delivers energy to the Host, unless earlier termination as specifically provided for in this Agreement. After the Initial Term, this Agreement shall continue for successive two (2) year periods ("Extended Terms"); provided, however, either party may terminate this Agreement by giving written notice ninety (90) days prior to the end of the Initial Term or any Extended Term, and thereafter the Agreement shall terminate at the end of the Term or such other date as the parties shall mutually establish. The Initial Term and Extended Term shall be referred to herein as the "Term."

3.6 **Right of Entry:**
The Host hereby grants to TPF a non-exclusive right of entry to the Facility during business hours (except as otherwise expressly permitted by the Host) for the purposes of installing, operating, maintaining, and repairing the System. TPF or its subcontractor(s) may enter the Facility during normal business hours for the purpose of making routine repairs and alterations or as requested by TPF and approved by the Host or at any time to make emergency repairs.

3.7 **Purchase and Sale of Electrical Power Output and Thermal Energy:**
The Host agrees to purchase from TPF the electrical and thermal energy produced by the System and consumed by the Facility. Usage of electrical and thermal energy shall be metered and measured by the same means applied to commercial suppliers.

Host shall purchase, and TPF shall sell and deliver, the thermal energy and electric power output of the System substantially as specified on Exhibit B attached hereto. **TPF** and Host agree that the published rate schedules used by the **[Local Electric Utility]** and the **[Local Gas Utility]** serving the Host as of the Effective Date shall be the basis for determining the price of the thermal energy and electric power produced by the System and used by the Host. Electric energy consumption shall be billed monthly to the Host at a **[20%]** reduction from the rate that the **[Local Electric Utility]** is charging on the Effective Date. Thermal energy used will be converted to gas consumption that would have been required in the absence of the System and shall be billed monthly to the Host at a **[20%]** reduction from the rate the **[Local Gas Utility]** is charging on the Effective Date. The efficiency of the water heater and heat exchangers in use by the Host on the Effective Date shall be taken from the faceplate on the equipment when converting BTUs delivered from the System into equivalent consumption for billing purposes.

TPF shall pay for all fuel needed to operate the System. TPF shall pay for all costs associated with, and necessary for, the measurement of the fuel used by the System.

TPF shall furnish to the Host monthly statements showing the amount of electrical and thermal. energy outputs of the System actually used by the Facility and the rates applicable thereto. The Host shall pay such invoices within twenty (20) days of presentment. Any portion of such invoice amount not paid within such twenty (20) day period shall bear interest at the annual rate of two and one-half percent (2.5%) over the Prime Rate announced from time to time by the Wall Street Journal (but not to exceed the maximum permitted by law).

TPF acknowledges that the Host may purchase supplemental electric and thermal energy from any other source that the Host desires, in its sole and absolute discretion. The Host may install such energy-savings devices, as it may desire, in its sole and absolute discretion.

3.8 **Alteration to Facility:**
The Host shall, at its sole cost, expense, and liability, maintain the Facility and pay all utilities, taxes, and bills associated therewith, except for those attributable to the System payable by TPF as set forth in this Agreement. The Host agrees to not undertake any structural alterations or repairs to the Facility which may adversely impact the operation and maintenance of the System (i) without giving prior written notice to TPF, setting forth the work to be undertaken (except in the event of emergency repairs, in which event notices may be given by telephone), and (ii) without offering TPF the opportunity to advise the Host in conducting the structural work in a manner that will not result in damage to, or adversely affect, the operation of the System. If the Host fails to provide prior written notice to TPF and receive TPF's advice concerning the structural work as set forth above, the Host shall be responsible for all damages resulting therefrom. The Host shall not be responsible for damages to the System or losses to TPF if the Host has given prior written notice and has followed TPF's advice with respect to conducting the structural work.

TPF shall, at its sole cost, expense, and liability, repair any damage to the Facility resulting from the installation and/or operation of the System, but shall have no other responsibility with respect to maintenance of the Facility, except with respect to such maintenance as may be directly attributable to or arise from the installation or operation of the System.

3.9 **Maintenance of the System:**
TPF shall, at its sole cost, expense, and liability, perform all routine, emergency repairs, maintenance, and operation of the System. TPF shall provide, at its sole cost, expense, and liability, all labor, material, and other supplies necessary to perform such maintenance, repair, or operation.

In the event of a partial or complete failure of the System, TPF agrees to respond within twenty-four (24) hours following written or telephone notification from the Host, and will effect such repairs as soon as reasonably possible to restore the System.

3.10 **Use of Subcontractors:**
TPF shall be permitted to use subcontractors to perform its obligations under this Agreement. However, TPF shall continue to be responsible for the quality of the work performed by its subcontractors as provided in Section 3.1.

3.11 **Damages and Termination:**
The Host may terminate this Agreement without cost or liability (i) as provided by specific provisions of this Agreement, or (ii) upon TPF material breach hereunder, after no less than thirty (30) days' prior written notice of the breach and a reasonable time for TPF to cure the breach. Upon such termination, TPF shall, at its sole cost, expense, and liability, remove the System. Said scope of work shall be limited to removal of the cogenerator and attached related equipment. The cement pad and all piping shall not be removed but the piping will be capped.

TPF may terminate this Agreement without cost or liability (i) upon Host's failure to make payment as provided in Section 3.7 of this Agreement, or (ii) upon the Host's material breach hereunder, after no less than thirty (30) days' prior written notice of the breach and a reasonable time for Host to cure the breach. TPF may pursue all other legal and equitable remedies available to it arising from such breach.

3.12 **Indemnification:**
TPF shall defend, indemnify, and hold the Host, its officers, directors, agents, employees, and contractors harmless from and against all liability and expenses (including reasonable attorney's fees and litigation costs) for property damage, or personal injury (including wrongful death) arising out of but only to the extent of the negligence or willful misconduct of TPF, or to the extent of the negligence or willful misconduct of its officers, employees, agents, and contractors in the design, construction, installation, maintenance, or operation of the System.

The Host agrees to defend, indemnify, and hold TPF, its officers, directors, agents, employees, and contractors harmless from and against all liability and expenses (including reasonable attorney's fees and litigation expenses) for property damage or personal injury (including wrongful death) arising out of or in any way related to the Host's negligence or willful misconduct, or the negligence or willful misconduct of its officers, directors, agents, employees, and contractors.

The parties' respective indemnification obligations shall survive the expiration or termination of this Agreement for the period set forth in the applicable statute of limitation.

3.13 **Limitations of Liability:**
Neither TPF nor its employees, its subcontractors or suppliers shall be liable for any indirect, special, incidental, exemplary, or consequential loss or damage of any nature arising out of their performance or non-performance hereunder. TPF and Host hereby specifically agree that Host is fully responsible for the upkeep and maintenance of all of Host's equipment that will utilize the System's outputs, such as, but not limited to, hot water, process heat, space heating, and absorption air conditioning. Host agrees to properly maintain in full working order any and all of Host's equipment that Host may shut down while utilizing the System. TPF shall specifically not be liable for any direct, indirect, special, incidental, exemplary, or consequential loss or damage of any nature arising out of Host's performance or nonperformance of Host's equipment, while the System is shut down for maintenance, repair or replacement.

In no event shall TPF liability arising out of or in connection with the performance or nonperformance of this Agreement or the design or installation of any equipment or the System exceed the total of monthly payments made by the Host under Section 3.7 of this Agreement as of the date of the claim. The provisions of this Section 3.13 shall apply whether such liability arises in contract, tort (including negligence), strict liability or otherwise. Any action against TPF must be brought within one (1) year after the cause of action accrues.

In no event shall TPF be responsible for any damages arising out of any failure to perform or delay due to any cause beyond TPF's reasonable control. In such event, TPF shall be entitled to an extension of time as necessary to overcome the cause of the failure to perform or delay.

3.14 **Assignment**:
This Agreement may not be assigned in whole or in part without the express written consent of the other party, which consent may not be unreasonably withheld or delayed. Notwithstanding the foregoing, TPF may assign, mortgage, pledge, or otherwise transfer its interests in this Agreement to (i) any lender, or (ii) an affiliate of TPF, in either case without obtaining the consent of Host.

3.15 **Arbitration**:
If a dispute arises under this Agreement, the parties shall promptly attempt in good faith to resolve the dispute by negotiation. The parties agree that any disputes or claims between TPF, and Host arising out of or relating to this Agreement, or the breach thereof, not settled by negotiation, shall be settled by arbitration, conducted in accordance with the Commercial Arbitration Rules of the American Arbitration Association in effect at that time, except as modified herein, at a location specified by TPF. All disputes shall be decided by a single arbitrator. A decision shall be rendered by the arbitrator no later than nine months after the demand for arbitration is filed, and the arbitrator shall state in writing the factual and legal basis for the award. The arbitrator shall issue a scheduling order that shall not be modified except by the mutual agreement of the parties. Any award rendered by the arbitrator shall be final, and judgment may be entered upon it in accordance with applicable law in any court having jurisdiction thereof. The prevailing party shall recover all costs, including attorney's fees, incurred as a result of the dispute.

3.16 **Governing Law:**
This Agreement shall be governed, interpreted, and enforced in accordance with the laws of the...state in which TPF is located.

3.17 **No Third Party Beneficiaries:**
Host and TPF expressly agree that they do not intend to benefit

any person or entity not a signatory to this Agreement. No third party beneficiaries are intended or shall be created by operation of this Agreement.

3.18 **Arms' Length Negotiations:**
This Agreement is the product of mutual arms' length negotiations in which both the Host and TPF have been represented by legal counsel. Accordingly, the rule of judicial construction that ambiguities in a document will be construed against the drafter of that document shall have no application to the interpretation or enforcement of this Agreement.

3.19 **Insurance:**
At all times during the Term of this Agreement, TPF shall maintain commercial general liability insurance in the amount of $1,000,000 per person, $2,000,000 aggregate per incident. **[Prior to installation of the System, TPF shall cause its insurer to issue endorsements to its policy of commercial general liability insurance, naming the Host and its employees, agents, and contractors as additional insured under said policy.]** TPF agrees that said policy shall not be canceled or reduced in coverage without thirty (30) days written notice to the Host. Insurance shall be provided on an "occurrence," not a claims made basis, and shall be primary and noncontributing with any insurance that the Host may elect to maintain.

3.20 **System Purchase Option:**
The Host shall have the right to purchase the System from TPF at a negotiated fair market value on each anniversary of this Agreement.

3.21 **Minimum Electric Energy Take:**
The Host agrees to accept a minimum amount of electric power output of the System not to fall below [60,000] kilowatt-hours (kWh) per month. In the event that the Host does not consume said amount of kWhs the monthly invoice from TPF shall state the required minimum at a rate of [$.073] per kWh.

3.22 **Termination Due to Unprofitable Operation:**
At any time during the Term of this Agreement should TPF deter-

mine in its sole opinion that the Facility is unprofitable, for whatever reason, then TPF may cease operations upon fifteen (15) days' written notice to the Host.

Section 4.   Access to Premises

4.1 **Easement:**
In consideration of the energy services provided by TPF under this Agreement, the Host hereby grants an easement ("Easement") to TPF on such portion of the Facility as needed for TPF's equipment and grants to TPF an access and maintenance easement appurtenant thereto as is reasonably necessary for construction, installation, maintenance, and operation of the System during the Term of this Agreement. As part of the Easement, thermal energy and electrical power output shall be transmitted from the System, and necessary lines connecting the System with the Host's pre-existing conventional electrical and heating/cooling systems shall be run in, under, over, across, and through the Facility. The term of the Easement conferred hereby shall be identical to the Term of this Agreement. The precise area of the Easement shall be determined mutually by Host and TPF. Where required by applicable State law, TPF, and Host shall provide for the Easement in a separate agreement or document and Host shall cooperate with TPF to take all steps necessary to execute and record such Easement in the public records.

4.2 **Maintenance, Alterations, and Repairs of Easement and Facility:**
The Host shall be obligated to maintain fully the Facility and the Easement as prescribed herein and to pay all utilities, taxes, and bills (other than those assumed by TPF attributable thereto. The Host shall not undertake any action that would in any manner adversely affect TPF's operation of the System. TPF shall repair any damage to the Easement resulting from installation or operation of the System, but shall have no responsibility with respect to maintenance of the Easement.

Section 5.   Miscellaneous

5.1 **Notices:**
Notices shall be given by (i) certified mail, postage paid, or (ii)

delivery services such as Federal Express or similar service, or (iii) by facsimile. Notice sent by certified mail shall be deemed received two days after deposit in the United States mail. All notices not given by mail shall be deemed received upon actual receipt to the person to whom the notice is directed. Notices shall be addressed as follows:

To the Host:

    Telephone:
    Facsimile:

To TPF:

    2020 National St.
    Anytown, USA

5.2 **Integration and Modification:**

This Agreement represents the entire understanding of the parties with respect to the matters herein discussed. It supersedes in its entirety all prior or contemporary oral or written agreements. This Agreement may be amended only by writing and subscribed by the parties hereto.

IN WITNESS WHEREOF, the parties hereto have hereunto set their hand the day and year first above written.

    Third-party Financier
    Any Street
    Anytown, USA

    _____

    Officer of TPF
    Its: Vice President

    Host: _____

    _____

    Printed Name, Signature and Title

Rev. 8/31/98

Appendix II

Energy Conversion Tables

ENERGY STANDARDS:

1 THERM	=	100,000 Btu
1 MCF* OF NATURAL GAS	=	10 THERMS
1 GALLON OF OIL	=	1.43 THERMS
1 GALLON OF PROPANE	=	0.925 THERMS
1 POUND OF COAL	=	12,000 Btu
1 KWH OF ELECTRICITY	=	3,413 Btu
1 POUND OF WOOD WASTE	=	8,000 Btu
1 POUND OF GARBAGE	=	4,000 Btu
1 POUND OF STEAM	=	1,175 Btu
BOILER EFFICIENCY	=	80%
WATER HEATER EFFICIENCY	=	70 to 85%

*MCF ONE THOUSAND CUBIC FEET

THERMAL EQUIVALENTS:

1 MCF OF NATURAL GAS	=	7 GALLONS OF OIL
1 MCF OF NATURAL GAS	=	293 KW
1 TON OF COAL	=	1.5 TONS OF WOOD
1 TON OF COAL	=	3 TONS OF GARBAGE
1 TON OF COAL	=	24 MCF OF NATURAL GAS
1 TON OF COAL	=	168 GALLONS OF OIL

PRIME MOVER DATA:

	GAS TURBINES	RECIPROCATING ENGINES
EFFICIENCY:	26%	40%

ELECTRICITY GENERATED FROM 1 MCF OF NATURAL GAS:

	76 Kwh	117 Kwh

STEAM RAISED FROM REMAINING ENERGY:

REMAINING Btu's:	740,000	600,000
STEAM PRODUCED	503 Pounds	184 Pounds

Appendix III

Heat Loss in Swimming Pools

HEAT LOSS IN THERMS PER SQUARE FOOT
OF POOL SURFACE AREA
vs.
MONTHS OF THE YEAR

CHART I: INLAND POOL ENERGY ANALYSIS. COVERED FROM 2300 HOURS (11:00 PM) TO 1100 HOURS (11:00 AM). HEATER ON FROM 1100 HOURS TO 2300 HOURS.
AVERAGE AMBIENT TEMPERATURE: 70 DEGREES F.

CHART II: COASTAL POOL ENERGY ANALYSIS. COVERED FROM 2300 HOURS TO 1100 HOURS. HEATER ON 24 HOURS PER DAY.
AVERAGE AMBIENT TEMPERATURE: 60 DEGREES F.

Appendices 171

Chart I. Inland Pool Energy Analysis—Cover on 2300-1100; Heater on 1100-2300

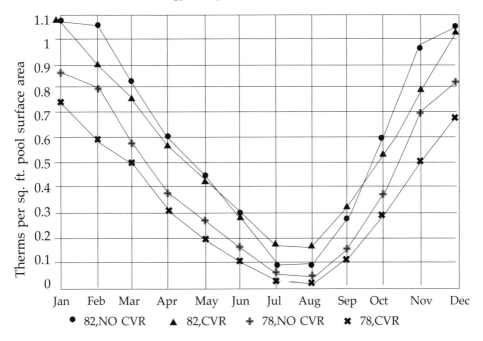

Chart II. Coastal Pool Energy Analysis; Cover on 2300-1100; Heater on 24 hr/day

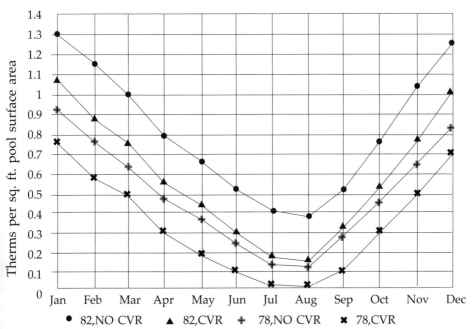

Appendix IV

Typical Small-scale Cogeneration Energy Production

SIZE UNIT	60 kW	75 kW	120 kW
FUEL INPUT Btu/hr	780,000	863,000	1,070,000
THERMAL OUTPUT Btu/hr	480,000	490,000	562,000
ELECTRICAL OUTPUT Btu/hr	204,750	246,000	409,500
THERMAL EFFICIENCY, %	87.89%	85.28%	90.79%
HOT WATER TEMP, DEG. F.	— UP TO 220°F —		
GPM FLOW RATE	25	28	22
WEIGHT, LBS.	3100	3200	3800
DIMENSIONS, INCHES	80L × 44W × 61H	82L × 44W × 46H	98L × 44W × 61H
NOISE LEVEL @ 6 FT.	70 dBA	75 dBA	72 dBA

Above figures are representative of a composite of manufacturers units and will vary with each specific manufacturer.

Appendix V

Typical Utility Rate Schedules— Gas and Electricity

The following tables are meant to portray the types of schedules a supplier of gas and electricity would publish demonstrate available rates for different types of service.

NATURAL GAS SCHEDULES:

Schedule GN-1: Commercial and Industrial Natural Gas Service
Schedule GN-2: Commercial and Industrial Natural Gas Service (Large Customer)
Schedule GTC: Natural Gas Transportation Service for Core Customers
Schedule GTCA: Natural Gas Transportation Service for Core Aggregation Customers
Schedule G-NGV: Sale of Natural Gas for Motor Vehicle Fuel
Schedule GT-NGV: Transportation of Customer-Owned Gas for Motor Vehicle Service
Schedule GPC: Gas Procurement for Core Customers
Schedule GCORE: Core Subscription Natural Gas Service for Retail NonCore Customers
Schedule GPNC: Gas Procurement for NonCore Customers
Schedule GPNC-S: Gas Procurement for NonCore Customers
Schedule GP-SUR: Customer Procured Gas Franchise Fee Surcharge

ELECTRIC RATE SCHEDULES:

Schedule A: General Service
Schedule AD: General Service - Demand Metered
Schedule AL-TOU: General Service - Large - Time Metered

San Diego Gas & Electric Company
San Diego, California

Revised Cal. P.U.C. Sheet No. 9453-G
Canceling Revised Cal. P.U.C. Sheet No. 9373-G

Sheet 1 of 2

SCHEDULE GN-1
COMMERCIAL AND INDUSTRIAL NATURAL GAS SERVICE

APPLICABILITY

Applicable to natural gas service classified as non-residential core service, and assigned end-use priorities P-1, P-2A, or P-2B, where the average use per facility does not exceed 20,800 therms per month during the same season gas is used. End-use priorities are defined in Rule 14.

Non-profit group living facilities taking service under this schedule may be eligible for a 15% low-income rate discount on their bill, if such facilities qualify to receive service under the terms and conditions of Schedule G-LI.

Agricultural Employee Housing Facilities, as defined in Schedule G-LI, may qualify for a 15% CARE discount on the bill if all eligibility criteria set forth in Form 142-4032 is met.

TERRITORY

Within the entire territory served natural gas by the utility.

RATES

	Per Meter Per Month	
Customer Charge	$ 5.00	
Commodity Charges:		
Summer Usage*, $ per therm:		
0 - 1,000 therms	$0.74737	I
All Excess	$0.51897	I
Winter Usage*, $ per therm:		
0 - 1,000 therms	$0.86311	I
All Excess	$0.52564	I

* Summer Usage occurs from April 1 to November 30.
  Winter Usage occurs from December 1 to March 31.

The number of therms shall be determined in accordance with the provisions of Rule 2.

The rates shown above include the Schedule GPC rate of $0.29120 per therm as a reference rate. These rates will be adjusted on a monthly basis as the Schedule GPC rate is adjusted.   I

Minimum Charge:
The minimum charge shall be the customer charge.

Franchise Fee Differential:
A franchise fee differential of 1.0% will be applied to the monthly billings calculated under this schedule for all customers within the corporate limits of the City of San Diego. Such franchise fee differential shall be so indicated and added as a separate item to bills rendered to such customers.

(Continued)

Advice Ltr. No.	1087-G	Issued by **William L. Reed** Vice President Regulatory Affairs	Date Filed	March 4, 1998
Decision No.			Effective	March 9, 1998
			Resolution No.	

San Diego Gas & Electric Company
San Diego, California

Revised Cal. P.U.C. Sheet No. 9454-G

Canceling Revised Cal. P.U.C. Sheet No. 9374-G

Sheet 1 of 2

SCHEDULE GN-2

COMMERCIAL AND INDUSTRIAL NATURAL GAS SERVICE

APPLICABILITY

Applicable to natural gas service classified as non-residential core service, and assigned end-use priorities P-1 or P-2A, where the average use per facility exceeds 20,800 therms per month during the same season gas is used. Customers eligible for service under this schedule may also elect transmission-only service as defined in Special Condition 2. End-use priorities are defined in Rule 14.

Non-profit group living facilities taking service under this schedule may be eligible for a 15% low-income rate discount on their bill, if such facilities qualify to receive service under the terms and conditions of Schedule G-LI.

Agricultural Employee Housing Facilities, as defined in Schedule G-LI, may qualify for a 15% CARE discount on the bill if all eligibility criteria set forth in Form 142-4032 is met.

TERRITORY

Within the entire territory served natural gas by the utility.

RATES

	Per Meter Per Month
Customer Charge	$ 75.00
Commodity Charges:	
Summer Usage*, $ per therm:	
0 - 6,000 therms	$0.62529 I
All Excess	$0.44906 I
Winter Usage*, $ per therm:	
0 - 6,000 therms	$0.75262 I
All Excess	$0.47462 I

* Summer Usage occurs from April 1 to November 30.
  Winter Usage occurs from December 1 to March 31.

The number of therms shall be determined in accordance with the provisions of Rule 2.

The rates shown above include the Schedule GPC rate of $0.29120 per therm I
as a reference rate. These rates will be adjusted on a monthly basis as
the Schedule GPC rate is adjusted.

Minimum Charge:
The minimum charge shall be the customer charge.

Franchise Fee Differential:
A franchise fee differential of 1.0% will be applied to the monthly billings calculated under this schedule for all customers within the corporate limits of the City of San Diego. Such franchise fee differential shall be so indicated and added as a separate item to bills rendered to such customers.

(Continued)

	Issued by	Date Filed	March 4, 1998
Advice Ltr. No. 1087-G	**William L. Reed**	Effective	March 9, 1998
Decision No.	Vice President Regulatory Affairs	Resolution No.	

SDG&E
An Enova Company
San Diego Gas & Electric Company
San Diego, California

Revised Cal. P.U.C. Sheet No. **9455-G**
Canceling Revised Cal. P.U.C. Sheet No. **9375-G**

Sheet 1 of 3

SCHEDULE GTC

NATURAL GAS TRANSPORTATION SERVICE FOR CORE CUSTOMERS

APPLICABILITY

Applicable to intrastate transportation-only natural gas service classified as:

(1) Core usage that is measured through a single gas meter with usage in excess of 250,000 therms annually or an average usage of 20,800 therms per month during the season gas is used; or

(2) Core usage that is measured through the same gas meter or located on the same premise as noncore usage receiving intrastate transportation-only service for the same customer of record.

Core customers who do not meet the above applicability may opt to aggregate their core loads with other core customers and receive service under Schedule GTCA. Service under this schedule must be taken in conjunction with service under Schedule GP-SUR.

Non-profit group living facilities taking service under this schedule may be eligible for a 15% low-income rate discount on their bill, if such facilities qualify to receive service under the terms and conditions of Schedule G-LI.

Agricultural Employee Housing Facilities, as defined in Schedule G-LI, may qualify for a 15% CARE discount on the bill if all eligibility criteria set forth in Form 142-4032 is met.

TERRITORY Applicable throughout the utility's service territory.

RATES

Otherwise Applicable Rate Schedule	Transported Volumes in Therms	Transport Rate $/Therm	
GR, GM, GS, GT: (Regular Rates)	Baseline Quantities 1/	$0.40617	R
	All Excess	$0.61868	R
GR-LI, GS, GT: (CARE Rates)	Baseline Quantities 1/	$0.30162	R
	All Excess	$0.48226	R
GN-1: Winter (Dec.1- Mar.31)	0 - 1,000 therms	$0.57234	R
	All Excess	$0.23487	R
Summer (Apr.1- Nov.30)	0 - 1,000 therms	$0.45660	R
	All Excess	$0.22820	R
GN-2: Winter (Dec.1- Mar.31)	0 - 6,000 therms	$0.46185	R
	All Excess	$0.18385	R
Summer (Apr.1- Nov.30)	0 - 6,000 therms	$0.33452	R
	All Excess	$0.15829	R
GT-NGV:	Uncompressed Gas	$0.16931	R

1/ See applicable residential rate schedule for baseline quantities.

(Continued)

Advice Ltr. No. 1087-G
Decision No.

Issued by
William L. Reed
Vice President
Regulatory Affairs

Date Filed **March 4, 1998**
Effective **March 9, 1998**
Resolution No.

Appendices

San Diego Gas & Electric Company
San Diego, California

	Revised Cal. P.U.C. Sheet No.	9456-G
Canceling	Revised Cal. P.U.C. Sheet No.	9229-G

Sheet 2 of 3

SCHEDULE GTC

RATES (Continued)

In addition to the rates shown above, all customers not exempted from the Rate Cap Transition Charge, shall have the Rate Cap Transition Charge of $0.03868 per therm added to their bill for service from August 1, 1997 through July 31, 1998. Customers that are exempted from the Rate Cap Transition Charge are those customers that procured their own gas on December 31, 1996. No other customers will be exempt from the Rate Cap Transition Charge. As a matter of clarification, the Rate Cap Transition Charge does not apply to customers, who on December 31, 1996, were solely transport customers, and who did not, on such date, purchase gas from the utility.

Standby Service Fee, per decatherm .... $ 10

>This fee shall be assessed to customers only during curtailments of transportation services to firm noncore customers. This fee will apply only to the difference between the customer's nominations and their confirmed deliveries.
>
>The customer's storage volumes, if available, may be used to offset the standby service fee. Revenues collected from this fee shall be credited to the utility's core gas fixed cost account (CFCA). Curtailments of standby services provided to core customers are described in Rule 14.
>
>Customers who receive service under this schedule shall also be eligible for standby services ahead of such services offered to noncore customers, including core subscription customers.

Billing adjustments may be necessary to reflect changes in volumes used in developing prior periods' transportation charges.

The minimum charge shall be the monthly customer charge, if applicable, of the otherwise applicable rate schedule for utility gas sales.

Self procuring customers may be subject to a Wheeler Ridge Access Fee.

All rates shown above include a core interstate transition cost surcharge (CITCS) of $0.00043 per therm.   R

The number of therms to be billed shall be determined in accordance with Rule 2.

A late payment charge may be added to the customer's billing charges whenever the customer fails to pay for services rendered under this schedule. See Rule 32 for further details.

Submetering of service under this schedule will be rendered in accordance with Rule 19.

Franchise Fee Differential

A franchise fee differential of 1% will be applied to the monthly billings calculated under this schedule for all customers within the corporate limits of the City of San Diego. Such franchise fee differential shall be so indicated and added as a separate item to bills rendered to such customers.

(Continued)

Advice Ltr. No.	1087-G	Issued by	Date Filed	March 4, 1998
		William L. Reed	Effective	March 9, 1998
Decision No.		Vice President Regulatory Affairs	Resolution No.	

San Diego Gas & Electric Company
San Diego, California

Revised Cal. P.U.C. Sheet No. 9457-G
Canceling Revised Cal. P.U.C. Sheet No. 9376-G

Sheet 1 of 4

SCHEDULE GTCA

NATURAL GAS TRANSPORTATION SERVICE FOR CORE AGGREGATION CUSTOMERS

APPLICABILITY

Applicable to intrastate transportation-only natural gas service classified as firm transportation of natural gas, with aggregated core usage having a minimum transport volume of 250,000 therms annually from a group of end-users, where each end-use meter is classified as core usage and located within SDG&E's service territory. Service under this schedule must be taken in conjunction with service under Schedule GP-SUR.

Non-profit group living facilities taking service under this schedule may be eligible for a 15% low-income rate discount on their bill, if such facilities qualify to receive service under the terms and conditions of Schedule G-LI.

Agricultural Employee Housing Facilities, as defined in Schedule G-LI, may qualify for a 15% CARE discount on the bill if all eligibility criteria set forth in Form 142-4032 is met.

TERRITORY

Applicable throughout the utility's service territory.

RATES

Otherwise Applicable Rate Schedule	Transported Volumes in Therms	Transport Rate $/Therm	
GR, GM, GS, GT: (Regular Rates)	Baseline Quantities 1/	$0.40617	R
	All Excess	$0.61868	R
GR-LI, GS, GT: (CARE Rates)	Baseline Quantities 1/	$0.30162	R
	All Excess	$0.48226	R
GN-1: Winter (Dec.1- Mar.31)	0 - 1,000 therms	$0.57234	R
	All Excess	$0.23487	R
Summer (Apr.1- Nov.30)	0 - 1,000 therms	$0.45660	R
	All Excess	$0.22820	R
GN-2: Winter (Dec.1- Mar.31)	0 - 6,000 therms	$0.46185	R
	All Excess	$0.18385	R
Summer (Apr.1- Nov.30)	0 - 6,000 therms	$0.33452	R
	All Excess	$0.15829	R
GT-NGV:	Uncompressed Gas	$0.16931	R

1/ See applicable residential rate schedule for baseline quantities.

(Continued)

Advice Ltr. No. 1087-G
Decision No.

Issued by
William L. Reed
Vice President
Regulatory Affairs

Date Filed March 4, 1998
Effective March 9, 1998
Resolution No.

San Diego Gas & Electric Company
San Diego, California

Revised Cal. P.U.C. Sheet No. 9458-G

Canceling Revised Cal. P.U.C. Sheet No. 9231-G

Sheet 2 of 4

SCHEDULE GTCA

RATES (Continued)

In addition to the rates shown above, all customers not exempted from the Rate Cap Transition Charge, shall have the Rate Cap Transition Charge of $0.03868 per therm added to their bill for service from August 1, 1997 through July 31, 1998. Customers that are exempted from the Rate Cap Transition Charge are those customers that procured their own gas on December 31, 1996. No other customers will be exempt from the Rate Cap Transition Charge. As a matter of clarification, the Rate Cap Transition Charge does not apply to customers, who on December 31, 1996, were solely transport customers, and who did not, on such date, purchase gas from the utility.

Standby Service Fee, per decatherm .... $ 10

> This fee shall be assessed to customers only during curtailments of transportation services to firm noncore customers. This fee will apply only to the difference between the customer's nominations and their confirmed deliveries.

> The customer's storage volumes, if available, may be used to offset the standby service fee. Revenues collected from this fee shall be credited to the utility's core gas fixed cost account (CFCA). Curtailments of standby services provided to core customers are described in Rule 14.

> Customers who receive service under this schedule shall also be eligible for standby services ahead of such services offered to noncore customers, including core subscription customers.

Billing adjustments may be necessary to reflect changes in volumes used in developing prior periods' transportation charges.

The minimum charge shall be the monthly customer charge, if applicable, of the otherwise applicable rate schedule for utility gas sales.

Self procuring customers may be subject to a Wheeler Ridge Access Fee.

All rates shown above include a core interstate transition cost surcharge (CITCS) of $0.00043 per therm.   R

The number of therms to be billed shall be determined in accordance with Rule 2.

A late payment charge may be added to the customer's billing charges whenever the customer fails to pay for services rendered under this schedule. See Rule 32 for further details.

Submetering of service under this schedule will be rendered in accordance with Rule 19.

Franchise Fee Differential

A franchise fee differential of 1% will be applied to the monthly billings calculated under this schedule for all customers within the corporate limits of the City of San Diego. Such franchise fee differential shall be so indicated and added as a separate item to bills rendered to such customers.

(Continued)

Advice Ltr. No. 1087-G	Issued by **William L. Reed** Vice President Regulatory Affairs	Date Filed Effective	March 4, 1998 March 9, 1998
Decision No.		Resolution No.	

San Diego Gas & Electric Company
San Diego, California

Revised Cal. P.U.C. Sheet No. 9459-G
Canceling Revised Cal. P.U.C. Sheet No. 9377-G

Sheet 1 of 2

SCHEDULE G-NGV

SALE OF NATURAL GAS FOR MOTOR-VEHICLE FUEL

APPLICABILITY

Applicable to natural gas service for motor-vehicle customers.

TERRITORY

Within the entire territory served natural gas by the utility. In the case of compressed natural gas service, available where capacity at existing refueling stations exists.

RATES

A. Compressed Natural Gas For all Other Fleets/Vehicles:
    Commodity Charge, cents per therm: .... 104.733   I

B. Compressed Natural Gas For Bus and Military Fleets:
    Commodity Charge, cents per therm: .... 77.621   I

C. Uncompressed Natural Gas For Motor-Vehicle Fleet Usage:
    Commodity Charge, cents per therm: .... 46.008   I

D. Co-Funded Stations:
    Commodity Charge, cents per therm: .... 75.371   I

The number of therms shall be determined in accordance with the provisions of Rule 2 - Description of Service.

Rates hereunder are subject to adjustment for any applicable taxes or surcharges which may be assessed.

Rates hereunder shall be revised annually, on the anniversary of the effective date of the Commission's Decision approving this tariff, to reflect the utility's Core Weighted Average Cost of Gas (Core WACOG) and the Energy Cost Adjustment Clause (ECAC) rate, in effect as of the anniversary date.

The rates shown above include the Schedule GPC rate of $0.29120 per therm as a reference rate. These   I
rates will be adjusted on a monthly basis as the Schedule GPC rate is adjusted.

Co-Funded Station Commodity Charge Proration

A Station jointly funded by the Customer and the utility, defined as a Co-funded Station, shall have the Commodity Charge prorated to reflect the Customer share of the funding for construction of the total station. The proration percentage used for the Commodity Charge shall be the same percentage as used in the co-funding of the total station. The Commodity Charge proration shall be between the Uncompressed Commodity Charge and the applicable Compressed Commodity Charge.

(Continued)

Advice Ltr. No. 1087-G

Issued by
William L. Reed
Vice President
Regulatory Affairs

Date Filed March 4, 1998
Effective   March 9, 1998

Decision No.   Resolution No.

Appendices 181

San Diego Gas & Electric Company
San Diego, California

Revised Cal. P.U.C. Sheet No. 9460-G
Canceling Revised Cal. P.U.C. Sheet No. 9233-G

Sheet 1 of 2

SCHEDULE GT-NGV
TRANSPORTATION OF CUSTOMER-OWNED GAS FOR MOTOR-VEHICLE SERVICE

APPLICABILITY

Applicable to the transportation of customer-owned gas for use in motor vehicles, to customers whose monthly gas throughput is at least 20,800 therms per month during the months in which gas is consumed.

Not applicable to a co-funded Station contract arrangement between the customer and the utility.

TERRITORY

Within the entire service territory served natural gas by the utility.

RATES

Uncompressed Natural Gas for Customer-funded Fueling Stations

Intrastate Transmission Charge (cents per therm).. 16.931   R

In addition to the rates shown above, all customers not exempted from the Rate Cap Transition Charge, shall have the Rate Cap Transition Charge of $0.03868 per therm added to their bill for service from August 1, 1997 through July 31, 1998. Customers that are exempted from the Rate Cap Transition Charge are those customers that procured their own gas on December 31, 1996. No other customers will be exempt from the Rate Cap Transition Charge. As a matter of clarification, the Rate Cap Transition Charge does not apply to customers, who on December 31, 1996, were solely transport customers, and who did not, on such date, purchase gas from the utility.

Compression of natural gas to the pressure required for its use as motor vehicle fuel will be performed by the customer using customer's equipment at the customer's designated premises.

The number of therms shall be determined in accordance with the provisions of Rule 2 - Description of Service.

Rates hereunder will be adjusted to reflect changes in the effective SoCalGas volumetric rate within 30 days of a change to such rates, and shall be subject to revision in conjunction with the utility's Biennial Cost Allocation Proceeding (BCAP).

Franchise Fee Differential

A franchise fee differential of 1% will be applied to the monthly billings calculated under this schedule for all customers within the corporate limits of the City of San Diego. Such franchise fee differential shall be so indicated and added as a separate item to bills rendered to such customers.

(Continued)

Advice Ltr. No. 1087-G	Issued by	Date Filed	March 4, 1998
	William L. Reed	Effective	March 9, 1998
Decision No.	Vice President Regulatory Affairs	Resolution No.	

San Diego Gas & Electric Company
San Diego, California

Revised Cal. P.U.C. Sheet No. 9461-G
Canceling Revised Cal. P.U.C. Sheet No. 9378-G

Sheet 1 of 2

SCHEDULE GPC

GAS PROCUREMENT FOR CORE CUSTOMERS

APPLICABILITY

Applicable to core customers who elect to purchase natural gas from the utility, including the benefits derived from utility-managed storage service.

TERRITORY

Within the territory where natural gas is provided by the utility.

RATES

Procurement Charges	$ per Therm	
Cost of Gas	0.23158	I
Capacity Charge 1/	0.01506	
FF&U	0.00493	I
Subtotal	0.25157	I
plus, Brokerage Fees (w/FF&U)	0.00095	
Subtotal	0.25252	I
plus, Rate Cap Transition Charge (w/FF&U) 2/	0.03868	
Total GPC Rate	0.29120	I

1/ Includes a core interstate transition cost surcharge (CITCS) of $0.00043 per therm.   R

2/ The Rate Cap Transition Charge shall be in effect from August 1, 1997 through July 31, 1998. Customers that were self-procurement customers on December 31, 1996 will be exempt from paying the Rate Cap Transition Charge.

The Total GPC rate above shall be subject to change no more than twice per month upon five (5) days written notice to reflect changes in the per unit charges listed above.

The rates contained in this schedule shall be included in each tariff where referenced.

Add any applicable taxes, fees, regulatory surcharges, or additional procurement charges imposed on the utility as a result of gas procurement under this schedule.

The number of therms will be determined in accordance with the provisions of Rule 2.

(Continued)

Advice Ltr. No.	1087-G	Issued by **William L. Reed** Vice President Regulatory Affairs	Date Filed: March 4, 1998
Decision No.			Effective: March 9, 1998
			Resolution No.

San Diego Gas & Electric Company
San Diego, California

Revised Cal. P.U.C. Sheet No. 9462-G

Canceling Revised Cal. P.U.C. Sheet No. 9379-G

Sheet 1 of 4

SCHEDULE GCORE

CORE SUBSCRIPTION NATURAL GAS SERVICE FOR RETAIL NONCORE CUSTOMERS

APPLICABILITY

Applicable to bundled natural gas service classified as firm purchase and transportation of natural gas; and

1. Cogeneration, regardless of size, whose facilities meet the efficiency standards specified in Section 218.5 (a) and (b) of the California Public Utilities Code.

2. Non-cogeneration gas service, where the annual average monthly use equals or exceeds 20,800 therms through a single meter and the customer has elected the noncore service classification. Customers with average monthly usage below 20,800 therms who were classified as noncore prior to September 17, 1993, may retain their noncore eligibility.

TERRITORY

Applicable throughout the utility's service territory.

RATES

MONTHLY CHARGES	UNITS	NON-COGEN	COGEN
Customer Charges (therms)			
0 to 3,000	$/meter /month of avg. demand	$16.00	$23.00
3,001 to 7,000		$83.00	$123.00
7,001 to 23,000		$151.00	$225.00
23,001 to 126,000		$303.00	$450.00
126,001 to 1,000,000		$608.00	$900.00
Over 1,000,000		$1,290.00	$1,911.00
Special Metering Fee	$/meter /month	$100.00	$100.00

VOLUMETRIC CHARGES	UNITS	NON-COGEN Medium Pressure	NON-COGEN High Pressure	COGEN non-Transm	Retail Noncore Transm
Volumetric Rate					
Winter (Dec - Mar)	$/therm	$0.34680	$0.30972	$0.30570	$0.29830
Summer (Apr - Nov)	$/therm	$0.32459	$0.29310	$0.29071	$0.28481

(Continued)

Advice Ltr. No. 1087-G	Issued by **William L. Reed** Vice President Regulatory Affairs	Date Filed Effective	March 4, 1998 March 9, 1998
Decision No.		Resolution No.	

SDG&E
San Diego Gas & Electric Company
San Diego, California

Revised Cal. P.U.C. Sheet No. 9463-G
Canceling Revised Cal. P.U.C. Sheet No. 9380-G

Sheet 1 of 3

SCHEDULE GPNC

GAS PROCUREMENT FOR NONCORE CUSTOMERS

APPLICABILITY

Applicable to noncore customers who elect natural gas procurement service from the utility. Gas services provided under this schedule shall include the purchase of gas supplies by the utility, the transportation of gas supplies to a point of interconnection with the utility's intrastate transportation system, and the benefits of utility-managed gas storage services. Service under this schedule shall be taken in conjunction with the applicable intrastate transportation service schedule(s).

TERRITORY

Within the territory where natural gas is provided by the utility.

RATES

	NON-UEG ($/therm)	UEG ($/therm)	
Cost of Gas	0.22158	0.22158	I
Capacity Charge	0.00391	0.00391	I
FF&U	0.00413	N/A	I
Subtotal	0.22962	0.22549	I
plus, Storage Service Fees	0.00126	0.00124	
plus, Brokerage Fee	0.00095	0.00093	
TOTAL GPNC RATE	0.23183	0.22766	I

The rates for a one-year purchase commitment shall be subject to change no more than twice per month upon five (5) days written notice to reflect changes in the per unit charges listed above.

Add any applicable taxes, fees, regulatory surcharges, or additional charges imposed on the utility as a result of providing gas procurement services under this schedule.

The number of therms will be determined in accordance with the provisions of Rule 2.

Franchise Fee Differential

A franchise fee differential of 1% will be applied to the monthly billings calculated under this schedule for all non-UEG customers within the corporate limits of the City of San Diego. Such franchise fee differential shall be so indicated and added as a separate item to bills rendered to such customers.

(Continued)

Advice Ltr. No. 1087-G
Decision No.

Issued by
William L. Reed
Vice President
Regulatory Affairs

Date Filed March 4, 1998
Effective March 9, 1998
Resolution No.

SDG&E
An Enova Company
San Diego Gas & Electric Company
San Diego, California

Revised Cal. P.U.C. Sheet No. 9464-G

Canceling Revised Cal. P.U.C. Sheet No. 9381-G

Sheet 1 of 3

SCHEDULE GPNC-S

GAS PROCUREMENT FOR NONCORE CUSTOMERS

APPLICABILITY

Applicable to noncore customers who elect natural gas procurement service from the utility. Gas services provided under this schedule shall include the purchase of gas supplies by the utility, and the transportation of gas supplies to a point of interconnection with the utility's intrastate transportation system. Service under this schedule shall be taken in conjunction with the applicable intrastate transportation service schedule(s).

TERRITORY

Within the territory where natural gas is provided by the utility.

RATES

	NON-UEG ($/therm)	UEG ($/therm)	
Cost of Gas	0.22158	0.22158	I
Capacity Charge	0.00391	0.00391	I
FF&U	0.00413	N/A	I
Subtotal	0.22962	0.22549	I
plus, Brokerage Fee (w/ FF&U)	0.00095	0.00093	
TOTAL GPNC-S RATE	0.23057	0.22642	I

The GPNC-S rate for a monthly purchase commitment shall be subject to change no more than twice per month upon five (5) days written notice to reflect changes in the per unit charges listed above.

Add any applicable taxes, fees, regulatory surcharges, or additional charges imposed on the utility as a result of providing gas procurement services under this schedule.

The number of therms will be determined in accordance with the provisions of Rule 2.

Franchise Fee Differential

A franchise fee differential of 1% will be applied to the monthly billings calculated under this schedule for all non-UEG customers within the corporate limits of the City of San Diego. Such franchise fee differential shall be so indicated and added as a separate item to bills rendered to such customers.

(Continued)

Advice Ltr. No.	1087-G	Issued by	Date Filed March 4, 1998
		William L. Reed	Effective March 9, 1998
Decision No.		Vice President Regulatory Affairs	Resolution No.

SDG&E
An Enova Company
San Diego Gas & Electric Company
San Diego, California

Revised Cal. P.U.C. Sheet No.	9465-G
Canceling Revised Cal. P.U.C. Sheet No.	9382-G

Sheet 1 of 2

SCHEDULE GP-SUR

CUSTOMER-PROCURED GAS FRANCHISE FEE SURCHARGE

APPLICABILITY

Applicable to all gas volumes procured by customers from an entity other than the utility and transported by the utility, pursuant to Senate Bill 278 (1993). These customers are generally referred to as "self-procurement" customers. Customers exempted from the surcharge by Senate Bill 278 (1993) are:

a. the State of California or a political subdivision thereof;

b. one gas utility transporting gas for end use in its Commission-designated service area through another utility's service area;

c. a utility transporting its own gas through its own gas transmission and distribution system, for purposes of generating electricity or for use in its own operations;

d. cogeneration customers, for that quantity of natural gas based on the Cogeneration Gas Allowance (CGA) in Schedule GTCG only for electricity sold to the utility.

TERRITORY

Applicable throughout the utility's gas pipeline system.

RATES

	$ per Therm	
	Core	Noncore
Gas Franchise Fee Surcharge (outside City of San Diego)	$0.00621	$0.00495	I
Gas Franchise Fee Surcharge (City of San Diego)	$0.00906	$0.00721	I

The Customer-Procured Gas Franchise Fee Surcharge, applicable to core and non-core customers, is comprised of the following components:

Core
a. The sum of the currently-defined Schedule GPC commodity rate, reduced for franchise fees; multiplied by,
b. The Franchise Fee Rate, as adopted in the utility's most recent General Rate Case decision.

Noncore
a. The currently-defined Schedule GPNC commodity rate reduced for franchise fees; multiplied by,
b. The Franchise Fee Rate, as adopted in the utility's most recent General Rate Case decision.

Schedule GP-SUR will be revised concurrently with each applicable Schedule GPNC rate revision.

(Continued)

Advice Ltr. No. 1087-G	Issued by **William L. Reed** Vice President Regulatory Affairs	Date Filed Effective Resolution No.	March 4, 1998 March 9, 1998
Decision No.			

SDG&E
San Diego Gas & Electric Company
San Diego, California

Revised Cal. P.U.C. Sheet No.   9569-E
Canceling   Revised Cal. P.U.C. Sheet No.   9470-E

Sheet 1 of 2

SCHEDULE A
GENERAL SERVICE

APPLICABILITY

Applicable to general service including lighting, appliances, heating, and power, or any combination thereof, and to three-phase residential service, including common use. This schedule is not applicable for single-phase service to residential customers. This schedule is not applicable to any customer whose Maximum Monthly Demand equals, exceeds, or is expected to equal or exceed 20 kW for 12 consecutive months. When demand metering is not available, the monthly consumption cannot equal or exceed 12,000 kWh per month for 12 consecutive months. This schedule is the utility's standard tariff for commercial customers with a demand less than 20 kW.

Non-profit group living facilities taking service under this schedule may be eligible for a 15% California Alternate Rates for Energy (CARE) discount on their bill, if such facilities qualify to receive service under the terms and conditions of Schedule E-LI.

Agricultural Employee Housing Facilities, as defined in Schedule E-LI, may qualify for a 15% CARE discount on the bill if all eligibility criteria set forth in Form 142-4032 is met.

TERRITORY

Within the entire territory served by the utility.

RATES

Basic Service Fee, per meter per month................. $7.77   I

Service Voltage	Secondary	Primary
Energy Charge per Kwh	$0.11378	$0.11028

    I

    Minimum Charge
       The minimum monthly charge shall be the Basic Service Fee.

    Energy Cost Adjustment
       An Energy Cost Adjustment, as specified in Section II.F. of the Preliminary Statement, will be included in each bill for service. The Energy Cost Adjustment amount shall be the product of the total kilowatt-hours for which the bill is rendered, multiplied by the Energy Cost Adjustment rate of $0.02547 per kWh. Base Energy Charges are included in the energy charges shown above less the Energy Cost Adjustment rate.

    Rate Cap Mechanism
       A Fuel Price Index Adjustment equal to the percentage as set forth on Schedule RCM shall be applied to the billings calculated under this schedule for all customers.

(Continued)

Advice Ltr. No. 1021-E	Issued by **William L. Reed** Vice President Regulatory Affairs	Date Filed January 24, 1997 Effective February 1, 1997 Resolution No.	
Decision No.			

San Diego Gas & Electric Company
San Diego, California

	Revised	Cal. P.U.C. Sheet No.	9405-E
	Revised		8994-E
Cancelling	Revised	Cal. P.U.C. Sheet No.	8560-E

Sheet 2 of 2

SCHEDULE A

RATES (Continued)   L

Franchise Fee Differential
Franchise fee differential of 1.9% will be applied to the monthly billings calculated under this schedule for all customers within the corporate limits of the City of San Diego. Such franchise fee differential shall be so indicated and added as a separate item to bills rendered to such customers.   L

SPECIAL CONDITIONS

1. <u>Definitions</u>: The Definitions of terms used in this schedule are found either herein or in Rule 1.

2. <u>Voltage</u>: Service under this schedule normally will be supplied at a standard available Voltage in accordance with Rule 2.

3. <u>Voltage Regulators</u>: Voltage Regulators, if required by the customer, shall be furnished, installed, owned, and maintained by the customer.

4. <u>Reconnection Charge</u>: In the event that a customer terminates service under this schedule and re-initiates service under this or any other schedule at the same location within 12 months, there will be a Reconnection Charge equal to the minimum charge which would have been billed had the customer not terminated service.

5. <u>Service to X-ray and Electronic Equipment</u>. Service under this schedule will be supplied to X-ray or Electronic Equipment, provided the apparatus is served from transformer capacity required to serve other general service load. In case the customer requests the utility to install excess transformer capacity to serve X-ray or electronic load, the customer charge will be increased by $1.00 per kVa of transformer capacity requested.

6. <u>Parallel Generation Limitation</u>. This schedule is not applicable to standby, auxiliary service, or service operated in parallel with a customer's generating plant, except as specified in Rule 1 under the definition of Parallel Generation Limitation.

7. <u>Compliance with Applicability Provisions</u>. For customers who are demand metered, applicability will be measured by the "20 kW for twelve consecutive months" provision. Applicability will be measured by the "12,000 kWh for twelve consecutive months" provision when demand metering is not available.

(Continued)

Advice Ltr. No. 998-E-A	Issued by	Date Filed December 23, 1996
Decision No. 96-12-077	**WILLIAM L. REED**	Effective December 23, 1996
	Vice President	Resolution No.
	Regulatory Affairs	

SDG&E
An Enova Company
San Diego Gas & Electric Company
San Diego, California

	Revised Cal. P.U.C. Sheet No.	9571-E
Canceling	Revised Cal. P.U.C. Sheet No.	9472-E

Sheet 1 of 3

SCHEDULE AD

GENERAL SERVICE - DEMAND METERED

APPLICABILITY

Applicable to general service including lighting, appliances, heating, and power, or any combination thereof to customers who have received service on this schedule on June 30, 1987. This schedule is not applicable to customers who request service after June 30, 1987, or who qualify for a baseline allowance. This schedule is not applicable to customers whose Maximum Monthly Demand has been less than 20 kW for three consecutive months, or whose Maximum Monthly Demand exceeds 500 kW for three consecutive months. Customers who discontinue service under this schedule cannot subsequently return to Schedule AD.

Non-profit group living facilities taking service under this schedule may be eligible for a 15% California Alternate Rates for Energy (CARE) discount on their bill, if such facilities qualify to receive service under the terms and conditions of Schedule E-LI.

Agricultural Employee Housing Facilities, as defined in Schedule E-LI, may qualify for a 15% CARE discount on the bill if all eligibility criteria set forth in Form 142-4032 is met.

TERRITORY

Within the entire territory served by the utility.

RATES

Service Voltage	Secondary	Primary	
Basic Service Fee per meter per month	$20.72	$20.72	I
Demand Charge, per kW Maximum Monthly Demand	$ 9.54	$ 9.16	I
Power Factor per kvar	$ .22	$ .22	
Energy Charge per Kwh	$0.08480	$0.08246	I

Energy Cost Adjustment
An Energy Cost Adjustment, as specified in Section II.F. of the Preliminary Statement, will be included in each bill for service. The Energy Cost Adjustment amount shall be the product of the total kilowatt-hours for which the bill is rendered, multiplied by the Energy Cost Adjustment rate of $0.02547 per kWh. Base Energy Charges are included in the energy charges shown above less the Energy Cost Adjustment rate.

(Continued)

	Issued by	Date Filed	January 24, 1997
Advice Ltr. No. 1021-E	**William L. Reed**	Effective	February 1, 1997
Decision No.	Vice President Regulatory Affairs	Resolution No.	

San Diego Gas & Electric Company
San Diego, California

Revised Cal. P.U.C. Sheet No. 9572-E
Canceling Revised Cal. P.U.C. Sheet No. 9473-E

Sheet 2 of 3

SCHEDULE AD

Rates (Continued)

Rate Cap Mechanism
A Fuel Price Index Adjustment equal to the percentage as set forth on Schedule RCM shall be applied to the billings calculated under this schedule for all customers.

Franchise Fee Differential
A Franchise Fee Differential of 1.9% will be applied to the monthly billings calculated under this schedule for all customers within the corporate limits of the City of San Diego. Such Franchise Fee Differential shall be so indicated and added as a separate item to bills rendered to such customers.

SPECIAL CONDITIONS

1. **Definitions**: The Definitions of terms used in this schedule are found either herein or in Rule 1.

2. **Voltage**: Service under this schedule normally will be supplied at a standard available Voltage in accordance with Rule 2.

3. **Voltage Regulators**: Voltage Regulators, if required by the customer, shall be furnished, installed, owned, and maintained by the customer.

4. **Reconnection Charge**: In the event that a customer terminates service under this schedule and re-initiates service under this or any other schedule at the same location within 12 months, there will be a Reconnection Charge equal to the minimum charge which would have been billed had the customer not terminated service.

5. **Power Factor**: The Power Factor rate shall apply to those customers that have a Power Factor Test Failure and will be based on the Maximum Kilovar Billing Demand. Those customers that have a Power Factor Test Failure will be required to pay for the Power Factor Metering that the utility will install.

6. **On-Peak Rate Limiter**: The On-Peak Rate Limiter only applies to customers taking service in conjunction with Schedules S or S-I. If, on a per kWh basis, the total charge for peak period demand and energy exceeds $0.83 per kWh in the summer or $0.32 per kWh in the winter, the bill for that service will be reduced such that the applicable $0.83 or $0.32 per kWh limit is not exceeded. This limiter only applies to energy and demand taken as backup service.

(Continued)

Advice Ltr. No. 1021-E
Decision No.

Issued by
William L. Reed
Vice President
Regulatory Affairs

Date Filed January 24, 1997
Effective February 1, 19
Resolution No.

		Revised Cal. P.U.C. Sheet No.	9573-E
San Diego Gas & Electric Company San Diego, California	Canceling	Revised Cal. P.U.C. Sheet No.	9474-E

SCHEDULE AD

Sheet 3 of 3

SPECIAL CONDITIONS (Continued)

6. <u>On-Peak Rate Limiter</u> (Continued)

   When a customer takes service in conjunction with Schedules S or S-I, a calculation is made in order to determine what demand and usage is subject to the $0.83 or $0.32 per kWh on-peak limiter and what is subject to the $5.18 average rate limiter. If a standby customer has a forced outage, that is demonstrated to the reasonable satisfaction of the utility within 60 days of occurrence, the on-peak demand and energy associated with the contracted standby kW are subject to the On-Peak Rate Limiter. All demand and usage not subject to the On-Peak Rate Limiter is subject to the average rate limiter . I I

7. <u>Service to X-ray and Electronic Equipment</u>. Service under this schedule will be supplied to X-ray or Electronic Equipment, provided the apparatus is served from transformer capacity required to serve other general service load. In case the customer requests the utility to install excess transformer capacity to serve X-ray or electronic load, the demand charge will be increased by $1.00 per kVa of transformer capacity requested.

8. <u>Parallel Generation Limitation</u>. This schedule is not applicable to standby, auxiliary service, or service operated in parallel with a customer's generating plant, except as specified in Rule 1 under the definition of Parallel Generation Limitation.

	Issued by	Date Filed	January 24, 1997
Advice Ltr. No. 1021-E	**William L. Reed**	Effective	February 1, 1997
Decision No.	Vice President Regulatory Affairs	Resolution No.	

San Diego Gas & Electric Company
San Diego, California

Revised Cal. P.U.C. Sheet No. 9574-E
Canceling Revised Cal. P.U.C. Sheet No. 9475-E

Sheet 1 of 5

SCHEDULE AL-TOU

GENERAL SERVICE - LARGE - TIME METERED

APPLICABILITY

Applicable to all metered non-residential customers whose monthly maximum demand equals, exceeds, or is expected to equal or exceed 20 kW. This schedule is optionally available to three-phase residential service, including common use, and to metered non-residential customers whose Monthly Maximum Demand is less than 20 kW. Any customer whose Maximum Monthly Demand has fallen below 20 kW for three consecutive months may, at their option, elect to continue service under this schedule or be served under any other applicable schedule. This schedule is the utility's standard tariff for commercial and industrial customers with a Monthly Maximum Demand equaling or exceeding 20 kW.

Non-profit group living facilities taking service under this schedule may be eligible for a 15% California Alternate Rates for Energy (CARE) discount on their bill, if such facilities qualify to receive service under the terms and conditions of Schedule E-LI.

Agricultural Employee Housing Facilities, as defined in Schedule E-LI, may qualify for a 15% CARE discount on the bill if all eligibility criteria set forth in Form 142-4032 is met.

TERRITORY

Within the entire territory served by the utility.

RATES

Service Voltage	Secondary	Primary	Primary Substation	Transmission	
Basic Service Fees per Meter per Month:					
Zero to 500 kW.........	$ 43.50	$ 43.50	$20,715.80	$ 43.50	I
500.1 to 10,000 kW......	$174.01	$174.01	$20,715.80	$174.01	I
Greater than 10,000 kW..	$174.01	$174.01	$31,073.70	$174.01	I
Distance Adjustment Fee per Foot per Month:	----	----	2.80	----	I
Demand Charges, per kW:					
Non-Coincident	4.68 I	4.26 I	0.41	0.41	
On-Peak Period					
Summer	20.52	20.00	14.67	14.57	I
Winter	4.76	4.64	2.99	2.97	I
Power Factor, per kvar	0.22	0.22	0.22	0.22	
Optional Contract					
Minimum Demand	0.52	0.52	0.52	0.52	
Energy Charge per kWh:					
Signaled Period 1G	0.10510	0.10309	0.10051	0.10003	
On-Peak					
Summer	0.07921	0.07720	0.07462	0.07413	
Winter	0.06631	0.06463	0.06248	0.06206	

(Continued)

Advice Ltr. No. 1021-E	Issued by **William L. Reed** Vice President Regulatory Affairs	Date Filed January 24, 1997 Effective February 1, 1997 Resolution No.
Decision No.		

Appendices 193

San Diego Gas & Electric Company
San Diego, California

Revised Cal. P.U.C. Sheet No. 9575-E
Canceling Revised Cal. P.U.C. Sheet No. 9476-E

Sheet 2 of 5

SCHEDULE AL-TOU

RATES (Continued)

Service Voltage	Secondary	Primary	Primary Substation	Transmission	
Semi-Peak					
Summer	0.04627	0.04523	0.04394	0.04365	I
Winter	0.04648	0.04544	0.04417	0.04388	I
Off-Peak					
Summer	0.03637	0.03580	0.03513	0.03491	I
Winter	0.03676	0.03618	0.03552	0.03529	I

Time Periods
All time periods listed are applicable to local time and exclude signaled periods. The definition of time will be based upon the date service is rendered.

	Summer May 1 - Sept 30	Winter All Other
On-Peak	11 a.m. - 6 p.m. Weekdays	5 p.m. - 8 p.m. Weekdays
Semi-Peak	6 a.m. - 11 a.m. Weekdays	6 a.m. - 5 p.m. Weekdays
	6 p.m. - 10 p.m. Weekdays	8 p.m. - 10 p.m. Weekdays
Off-Peak	10 p.m. - 6 a.m. Weekdays Plus Weekends & Holidays	10 p.m. - 6 a.m. Weekdays Plus Weekends & Holidays

Signaled Period 1G
Applicable to customers selecting the Optional Contact Closure Service: all hours during which the customer receives an electric contact closure from the utility.

Where the billing month contains time from both April and May or September and October, the on-peak period demand charges will be based on the demands registered in each month, weighted by the number of days billed in each month. Energy will be billed on the basis of the time period and season in which the usage occurred.

Non-Standard Seasonal Changeover
Customers may select on an optional basis to start the summer billing period on the first Monday of May and to start the winter billing period on the first Monday of October. Customers electing this option will be charged an additional $100 per year for metering equipment and programming.

Energy Cost Adjustment
An Energy Cost Adjustment, as specified in Section II.F. of the Preliminary Statement, will be included in each bill for service. The Energy Cost Adjustment amount shall be the product of the total kilowatt-hours for which the bill is rendered, multiplied by the Energy Cost Adjustment rate of $0.02547 per kWh. Base Energy Charges are included in the energy charges shown above less the Energy Cost Adjustment rate.

Rate Cap Mechanism
A Fuel Price Index Adjustment equal to the percentage as set forth on Schedule RCM shall be applied to the billings calculated under this schedule for all customers.

(Continued)

Advice Ltr. No. 1021-E
Decision No.

Issued by
William L. Reed
Vice President
Regulatory Affairs

Date Filed January 24, 1997
Effective February 1, 1997
Resolution No.

SDG&E
An Enova Company
San Diego Gas & Electric Company
San Diego, California

	Revised Cal. P.U.C. Sheet No.	9576-E
Canceling	Revised Cal. P.U.C. Sheet No.	9477-E

SCHEDULE AL-TOU

Sheet 3 of 5

Rates (Continued)

Franchise Fee Differential
A Franchise Fee Differential of 1.9% will be applied to the monthly billings calculated under this schedule for all customers within the corporate limits of the City of San Diego. Such Franchise Fee Differential shall be so indicated and added as a separate item to bills rendered to such customers.

SPECIAL CONDITIONS

1. **Definitions:** The Definitions of terms used in this schedule are found either herein or in Rule 1.

2. **Voltage:** Service under this schedule normally will be supplied at a standard available Voltage in accordance with Rule 2.

3. **Voltage Regulators:** Voltage Regulators, if required by the customer, shall be furnished, installed, owned, and maintained by the customer.

4. **Reconnection Charge:** In the event that a customer terminates service under this schedule and re-initiates service under this or any other schedule at the same location within 12 months, there will be a Reconnection Charge equal to the minimum charge which would have been billed had the customer not terminated service.

5. **Non-Coincident Demand Charge:** The Non-Coincident Demand Charge shall be based on the higher of the Maximum Monthly Demand or 50% of the Maximum Annual Demand.

6. **On-Peak Period Demand Charge:** The On-Peak Period Demand Charge shall be based on the Maximum On-Peak Period Demand.

7. **Power Factor:** The Power Factor rate shall apply to those customers that have a Power Factor Test Failure and will be based on the Maximum Kilovar billing demand. Those customers that have a Power Factor Test Failure will be required to pay for the Power Factor Metering that the utility will install.

8. **Average Rate Limiter:** The average rate for service under this schedule will be limited to $5.18 per kWh for all charges excluding the service charge and standby charges (if any). If the total billed energy and demand charges, on a per kWh basis, exceed $5.18 per kWh then $5.18 per kWh is substituted for that portion of the bill.

9. **On-Peak Rate Limiter:** The On-Peak Rate Limiter only applies to customers taking service in conjunction with Schedules S or S-I. If, on a per kWh basis, the total charge for peak period demand and energy exceeds $0.83 per kWh in the summer or $0.32 per kWh in the winter, the bill for that service will be reduced such that the applicable $0.83 or $0.32 per kWh limit is not exceeded. This limiter only applies to energy and demand taken as backup service.

(Continued)

Advice Ltr. No. 1021-E	Issued by **William L. Reed** Vice President Regulatory Affairs	Date Filed January 24, 1997 Effective February 1, 1997 Resolution No.
Decision No		

San Diego Gas & Electric Company
San Diego, California

Revised Cal. P.U.C. Sheet No. 9577-E
Canceling Revised Cal. P.U.C. Sheet No. 9478-E

Sheet 4 of 5

SCHEDULE AL-TOU

SPECIAL CONDITIONS (Continued)

9. When a customer takes service in conjunction with Schedules S or S-I, a calculation is made in order to determine what demand and usage is subject to the $0.83 or $0.32 per kWh on-peak limiter and what is subject to the $5.18 average rate limiter. If a standby customer has a forced outage, that is demonstrated to the reasonable satisfaction of the utility within 60 days of occurrence, the on-peak demand and energy associated with the contracted standby kW are subject to the On-Peak Rate Limiter. All demand and usage not subject to the On-Peak Rate Limiter is subject to the average rate limiter.

10. Parallel Generation Limitation. This schedule is not applicable to standby, auxiliary service or service operated in parallel with a customer's generating plant, except as specified in Rule 1 under the definition of Parallel Generation Limitation.

11. Seasonal Changeover Switching Limitation. Customers who elect the nonstandard Seasonal Changeover option of this schedule will be prohibited from switching service to the regular seasonal changeover for a 12-month period.

12. Limitation on Non-Standard Seasonal Changeover Availability. At the utility's sole option, the optional non-standard seasonal changeover provision is available to no more than ten additional Schedule AL-TOU and Schedule A6-TOU customers annually and; service will be provided in the order in which requests are received.

13. Terms of Optional Service. A customer receiving service under this schedule may elect to change to another applicable rate schedule, but only after receiving service on this schedule for at least 12 consecutive months. If a customer elects to discontinue service on this schedule, the customer will not be permitted to return to this schedule for a period of one year.

14. Basic Service Fee Determination. A customer whose Maximum Monthly Demand has exceeded 500 kW for three consecutive months in the most recent 12-month period is determined to be greater than 500 kW. Otherwise, a customer is determined to be less than or equal to 500 kW. A customer determined to be greater than 500 kW will be reassigned to less than or equal to 500 kW status only when the Maximum Monthly Demand remains less than or equal to 500 kW for 12 or more consecutive months. Differentiation of greater than or less than 10,000 kW will be based on the customer's Maximum Annual Demand.

(Continued)

Advice Ltr. No. 1021-E

Issued by
William L. Reed
Vice President
Regulatory Affairs

Date Filed January 24, 1997
Effective February 1, 1997

San Diego Gas & Electric Company
San Diego, California

	Revised	Cal. P.U.C. Sheet No.	9413-E
	Revised		9002-E
Cancelling	Revised	Cal. P.U.C. Sheet No.	9003-E

Sheet 5 of 5

SCHEDULE AL-TOU

SPECIAL CONDITIONS (Continued)

15. <u>Signaled Period 1G and Optional Contact Closure Service</u>. At the customer's option, in accordance with the provisions contained in utility's Request for Contact Closure Service (Form 142-1459), the utility will install, maintain, and operate, based on the conditions described herein, a device that will provide the customer with an electric contact closure indicating the start of a Signaled On-Peak Period, and this closure shall be opened at the end of each such period. The device will be located near the customers meter panel in space provided by the customer. In addition, the customer is entitled to receive telemetry service normally available to customers on Schedule A6-TOU, provided that the customer pays all additional expenses of this service.

    On-peak energy consumption during the billing period of the customer's choice will be billed at the Semi-Peak energy rate. Customers may designate the billing period under utility's Request for Service Form.

16. <u>Optional Contract Minimum Demand</u>. Optional Contract Minimum Demand shall be available to customers selecting the Optional Contact Closure Service. Optional Contract Minimum Demand shall be elected by the customer and recorded on the utility's Request for Contract Minimum Demand, (Form-1359). With 30 days written notice the customer may change this level by refiling a Request for Contract Minimum Demand, along with $25.00 plus the following charges: twelve minus one for each full month beyond the 30 day minimum notice given times any increase in demand, times the current Contract Minimum Demand rate (e.g. 13-month notice equals $25.00 charge regardless of contract demand level; 6-month notice equals $25 plus (7 times the increase in contract kW times the contract kW charge). The Signaled Period 1G Energy Charge will be applied to all kWh consumption provided in association with demand exceeding the Contract Minimum Demand. Consumption associated with demand not exceeding the Contract Minimum Demand will be billed at the energy charge of the pre-empted On-Peak, Semi-Peak or Off-Peak period. Under no circumstances shall a customer be permitted to sign for a Contract Minimum Demand in excess of his most current 12-month average demand at the time of filing the Request for Service. The Contract Minimum Demand will be equal to zero if the customer does not elect an Optional Contract Minimum Demand.

17. <u>Limitation of Availability</u>: Optional Contact Closure Service (Form 142-1459) shall be available subject to equipment availability.

Advice Ltr. No. 998-E-A
Decision No. 96-12-077

Issued by
WILLIAM L. REED
Vice President
Regulatory Services

Date Filed December 23, 1996
Effective December 23, 1996
Resolution No.

Appendix VI

Working A Cogeneration Project

A MODEL

These are the steps normal to finding, researching, analyzing, quoting, and implementing a typical cogeneration project. The same steps are utilized for a straight generation project as well (without heat recovery).

1. Making contact with the prospect via canvassing, following a lead, or using data bases to market advertise and receive responses.

2. Upon making contact, the first item is to determine what the prospect is currently using in the way of electricity and fuel to heat water, space heat, or other heat processes. This is best done by collecting the past twelve (12) months utility bills.

3. Determine the operating hours the equipment will be running. This is usually the hours the prospect's facility is open, although in some cases there may be significant electrical use even when the facility is closed, i.e. swimming pool heaters and pumps, security lighting, air conditioning, etc. Ask these questions.

4. Looking at the actual electric bills will show the rate structure under which the prospect's facility is on, the winter and summer energy and demand charges and any block pricing the utility imposes. Block pricing is when the first block of energy is priced differently than the second block, etc. When this occurs the analysis must take into account how much of the cheaper block is displaced and how much of the more expensive, first block, is displaced. Usually, 100% of block 2 will be displaced and less than 100% of block 1 will be displaced.

   Also, when the facility is not operating 24 hours/day, 7 days per week, the rate structure must be analyzed so that the proper cost per kWh and demand kW is used for the time the cogeneration equipment will be operating. Rate structures are such that the more expensive electricity is used during peak daylight hours in the summer and possibly different hours in the winter. Weekend power costs are usually the lowest, as are night time (10 PM to 6 AM) costs. Weighted averages must be calculated and used to make the cost of displaced electricity accurate.

5. Fuel bills, usually natural gas, must also be analyzed to determine how much of the fuel bill is used to heat water vs. those uses that may not be displaced by cogeneration, i.e. cooking, clothes drying, maybe space heating, etc. This is referred to as the "Thermal Load Factor" expressed as a % of the total gas usage.

6. Once the utility bills have been analyzed to produce the accurate input data, that data is then entered into a computer model to effect sizing the on-site generation and cogeneration equipment. Hours of operation of the equipment is also inputted. The optimum selection of equipment for grid connected generators is to not exceed the production of electricity vs. the consumption. A good rule of thumb is to keep the electrical production within 70 to 80 % of the consumption. That's to take into account the normal "spikes" that occur in a prospect's facility. Trying to displace 100% of the prospect's consumption will lead to overproduction of electricity during the "non-spike" periods and selling that overproduction back to the utility is not cost effective.

7. Once the generator size is selected, it behooves the analyzer to look at the beat side of the equation. The computer model will tell how many run hours it will take to displace the fuel used to beat water and that may be more or less than the hours in which the facility is operating. Keeping in mind that a Qualified Facility, one in which the PURPA law guarantees interconnection with the utility, requires a minimum efficiency to be attained, 42.5%, one must balance the beat side of the equation with the electrical side. Often this may require reducing the amount of on-site generation equipment to attain this minimum efficiency. However, in the face of de-regulation, the PURPA law will be phased out and the impact of Qualified Facility minimum efficiency will also phase out.

8. All of the above has led to determining an optimum size of generation equipment. The cost of installation must also be figured into the equation in order to give the prospect a true look at his overall savings and return on investment. Maintenance costs as well as operating costs, fuel consumption, are also factored in. Therefore, knowing the logistics of where the equipment can be installed vs. the location of the electric panel and the hot water heating equipment as well as the fuel supply, is important. Often, this can be determined by asking the prospect for his input and preparing an estimated proposal even before seeing the site.

9. If that estimated proposal appeals to the prospect, that's when a site visit can be made to "walk" the facility and determine from experience where the equipment can be located. Often, a contractor can be called in at this time to develop a firm proposal, but contractor's time is valuable and one should know that there is a high percentage opportunity that the project will sell before calling in a contractor. At this time in the proposition, it behooves the developer (salesman) to ask for a modest deposit.

10. The prospect now turns into a client, i.e. he has given you an order to proceed with the project. The events that follow take this route:
    a. Equipment is ordered from the manufacturer
    b. An order is issued to the installing contractor
    c. Installing contractor pulls a construction permit from the local city
    d. Application for interconnect is made to the local utility
    e. Air Quality permitting is addressed, usually a non-issue for equipment that is either exempt or demonstrates the ability to meet existing air pollution laws.
    f. Each city has It's own way of doing business and a plot plan, equipment cut sheets, electrical schematics, etc. may have to be produced. Some cities require a drawing prepared by a certified engineer which may add to the cost of the project.
    g. Each utility has it's own way of looking at the interconnect application. Protective devices must be proven to be in place to protect against events the utility determines "unsafe" should they occur. The equipment manufacturer must be prepared to show the utility its protective systems as well as Its single and three line electrical drawings.

11. The equipment is eventually installed and ready for start up. Coordination between the equipment vendor's service department and the manufacturer's service department is needed to insure a successful start up. Trouble shooting is the norm, and the system is tweaked to insure proper interface with the client's electrical and heating system.

12. The client is often interested in a maintenance agreement that may be one of the following:
    a. Preventative maintenance on a time and material basis, i.e. the service provider simply responds to planned needs or requested needs and charges based on the time spent and the parts supplied.
    b. Preventative maintenance on a contract basis, i.e. for a set cost per run hour or kWh produced, the service provider will respond to all planned maintenance events at no further cost to the client. All other costs, outside the manufacturer's warranty provisions, will be charged to the client on a time and material basis.
    c. Extended Maintenance and Warranty Agreement. This is where the service provider contracts with the client to provide all routine and emergency service at a fixed cost per run hour or kWh produced and will replace any part or component that requires replacement or repair at no additional cost to the client for as long as this contract is in place.

13. Ultimate goal: To have this client on the list of satisfied customers for use as a reference for future sales to new prospects.

COMMENTARY

Some other considerations that are often addressed during the process of defining a project and receiving an order are:

1. Financing. The prospect will often ask about leasing or other means of financing. Knowing who will lease this type of equipment is important and such companies as Balboa Capital and GE Capital are two examples of leasing companies that will respond to a lease requirement.

2. Tax Savings: The equipment can be depreciated over a 5 year period and lease payments, if structured as a true lease, may be written off as an expense, both deriving tax savings to a client.

3. Fuel costs: With the price of natural gas having risen to highest levels in decades, this is a major consideration. Even in cogeneration projects where a major portion of the client's natural gas usage to heat water is offset by the cogeneration system, the efficiency of the existing hot water system vs. that of the cogeneration system is higher therefore more fuel is consumed than is saved. Gas suppliers no longer look at cogeneration or self generation projects as "Non-Core" giving a reduced price for such gas usage.

4. Incubation time: The time from when a prospect allows you a look at his utility bills to the time an order is given is approximately 4-6 months on the average. The prospect is wary of investing to provide something that is already being provided via the utility company as well as his existing hot water system. Savings do not take on the same aspect as putting new production machinery in place or other capital investment projects. Yet, while a capital investment project may add revenue, since the average corporate profit is 5%, that's what the prospect may realize by those types of capital investments. Whereas, the cogeneration project may return him 30 to 40 percent return on his investment.

   New technology is the often heard by word. That makes references even more important to show the prospect the success of other projects where the clients are reaping the rewards of self and cogeneration.

5. Audits: After a project goes in it behooves the supplier to collect the client's records of generator operation, utility bills, maintenance records, etc. These will provide an audit of exactly what the self generation project has reaped. Usually a six month period should go by before such an audit takes place since the first couple of months of self generation operation may have been in tweaking the system to optimize its performance.

6. Incentives: Governments and utilities may offer incentives for cogeneration and self generation. These should be known and catego-

rized to the prospect to allow for maximum incentive for the prospect to take action. Presently, in California, the state offers low cost financing under the SAFEBIDCO program where 5% interest rate money is loaned for 5 years for projects that meet their standards. Also, the CPUC is offering incentives for cogeneration of $1/watt or 30% of the installation costs as a rebate to the client. Other incentives for biogas utilization, i.e. digester gas and landfill gas, are also being offered via the various state agencies. Recently, Southern California Gas Company has offered Technology Assistance programs in the form of cash to incentivize projects that use natural gas.

Index

A
absorber 125
absorber chiller 22, 23, 84, 97, 104, 105, 123, 124
Air Quality Management 51, 58
Athabasca region 6
average cost of electricity 38

B
BCHP 138
biomass 107
Bluepoint Energy Products 96
Bowman Power Systems 96, 115, 118, 119
Broad, USA 126, 127, 128

C
C. Itoh 90
CADER 144
California South Coast Air Quality 58
Capstone Turbine Corp 96, 112, 113, 114, 115, 116, 117, 128
casinos x
Caterpillar Co. 95
Coast Container Corp. 95
Coast Intelligen 96
Cogen Europe 151
cogeneration 1
Cogeneration Energy Services 88
cogeneration survey form 99, 100, 101
cogeneration system 2
computer program 38
condenser 125

conservation ix
Cummins 95

D
demand cost 26, 40
Denmark 150
deregulation 14, 15, 16
distributed generation 113, 129
distributed resources 130, 132
district heating 5
DOE 128, 141
DPCA 144
dump radiator 8, 48, 60

E
ecomomics 3, 35
economic analysis 102
Edison, Thomas 5
efficiency ix, 7
electricity offset 103
emissions 114
energy conversion tables 169
energy cost 25, 26, 43
environmental ix
EPRI 135, 141
ETL label 55
European Union 151
evaporator 125

F
FERC 7, 19, 48, 51, 52, 53, 56
FERC efficiency 7, 8, 27, 28, 34, 60, 141
finance xi
Finland 150

fitness center x, 21
five year cost/savings 39
food processors x, 22
fossil fuel values 11
fuel offset 103

G
Garret Corp. 111
gas rates—core/non-core 55
General Electric 95
geothermal 107
government funding 75
grants 75

H
Hanna, John 90, 91, 92, 93, 94
heat loss in swimming pools 170-171
heating cycle 125,126
Hess Microgen 96
Hitachi 126
Hitachi Building Equipment 87
hospitals x, 21
hotels x, 21
hybrid electric vehicles 114
hydroelectric 107

I
IEE Interconnection 141
Ingersoll-Rand Corp. 96, 115, 120, 121
instrumentation 61
Intelligen Products 95

K
kerosene 1
Kohler 115
kVA 70
kvar 70

L
Lawyer, Greg 92
leasing 73
Lumbert, Dave 87, 88, 90, 91

M
Mackay, Robin 111
maintenance 62, 85
maintenance contracts 63,64
metal plating 22
Microgen 89
microturbines; 96
monitoring 61
municipalities x, 21

N
NARUC 144
NCSL 144
Netherlands 150
Noe, Jim 111
NoMac 111, 113
North American Cogeneration 88, 89, 90
NO_x 58
nursing homes 22

O
OPEC 5, 6
operating cost 103

P
Pacific Gas & Electric 13, 18
power factor 69
PURPA 6,7, 8, 11, 12, 53, 68, 131

Qualified Facility 12, 19, 48, 51, 56, 57, 149

R
Raffesberger, Ray 95
Ratch, Herbert 87, 88, 90, 91, 93
Ratch, Margot 87, 88, 90, 91
rate schedule 41, 42,
recuperator 112
restaurants 21
return on investment 38

S
SAFE-BIDCO 75
San Diego Gas & Electric 13, 15, 16, 17, 18
shared savings 76, 77, 78 157-168
simple payback 103
solar energy 107
Solar Turbines 95
Southern California Gas Co., 128, 141
Southern California Edison 15, 18, 40
Sox 58
Standard Offer 12, 13
steam 23

T
Tecogen 96
thermal energy 47
thermal load fraction 98, 100
Thermex 87, 88
Thermo Electron Company 89
Trigen Energy Corp. 96
Turbec Americas, Inc. 96, 115, 119, 120

U
UL label 55
US Combined Heat & Power 137
Utility rate schedule 173-196

V
Value Chain 133
Volvo 115

W
wastewater treatment plants x
Waukesha 95
wind power 107
Working a cogeneration project 197-202

Y
Yazaki 126, 127
YMCA 80